Natural Resources and the Environment Series
Volume 9

Energy Alternatives in Latin America

**NATURAL RESOURCES AND THE ENVIRONMENT** Series

**SERIES DIRECTORS**
*Margaret R. Biswas and Asit K. Biswas*

**EDITORIAL BOARD**
*Essam El-Hinnawi, Nairobi; Huang Ping-Wei, Beijing; Mohammed Kassas, Cairo; Victor A. Kovda, Moscow; Walther Manshard, Freiburg; W. H. Matthews, Hawaii; M. S. Swaminathan, New Delhi.*

**Volume 1:** ENVIRONMENTAL IMPACTS OF PRODUCTION AND USE OF ENERGY
*Essam El-Hinnawi, United Nations Environment Programme*

**Volume 2:** RENEWABLE NATURAL RESOURCES AND THE ENVIRONMENT: PRESSING PROBLEMS IN THE DEVELOPING WORLD
*Kenneth Ruddle and Walther Manshard. United Nations University*

**Volume 3:** ASSESSING TROPICAL FOREST LANDS: THEIR SUITABILITY FOR SUSTAINABLE USES
*Richard Carpenter, Editor, East-West Center*

**Volume 4:** FUELWOOD AND RURAL ENERGY PRODUCTION AND SUPPLY IN THE HUMID TROPICS
*W. B. Morgan and R. P. Moss, United Nations University*

**Volume 5:** ECONOMIC APPROACHES TO NATURAL RESOURCE AND ENVIRONMENTAL QUALITY ANALYSIS
*Maynard M. Hufschmidt and E. Hyman, Editors, East-West Center*

**Volume 6:** RENEWABLE SOURCES OF ENERGY AND THE ENVIRONMENT
*Essam El-Hinnawi and Asit K. Biswas, Editors*

**Volume 7:** GLOBAL ENVIRONMENTAL ISSUES
United Nations Environment Programme, *Essam El-Hinnawi and Manzur-Ul-Haque Hashmi, Editors*

**Volume 8:** THE WORLD ENVIRONMENT 1972-1982
*Martin Holdgate, Mohammed Kassas and Gilbert White, Editors; Essam El-Hinnawi, Study Director, United Nations Environment Programme*

**Volume 9:** ENERGY ALTERNATIVES IN LATIN AMERICA
*Francisco Szekely, Editor. United Nations Environment Programme*

**Volume 10:** INTEGRATED PHYSICAL SOCIO-ECONOMIC AND ENVIRONMENTAL PLANNING
*Yusuf J. Ahmad and Frank G. Müller, Editors. United Nations Environment Programme*

**Volume 11:** THIRD WORLD AND THE ENVIRONMENT
*Essam El-Hinnawi and Asit K. Biswas, Editors*

**Volume 12:** DEVELOPMENT WITHOUT DESTRUCTION: EVOLVING ENVIRONMENTAL PERCEPTIONS
*Mostafa K. Tolba*

# Energy Alternatives in Latin America

*A Study sponsored by:*
United Nations Development Programme
United Nations Environment Programme
Latin American Energy Organization

*Editor*
Francisco Szekely

*Published by*
TYCOOLY INTERNATIONAL PUBLISHING LIMITED, DUBLIN

*All rights reserved. No part of this publication may be reproduced, stored in a retrieval system or transmitted, in any form or by any means, electronic, electrostatic, magnetic tape, mechanical, photocopying, recording or otherwise, without the prior permission in writing from the publishers.*

*Published by:*
Tycooly International Publishing Ltd.,
6 Crofton Terrace,
Dun Laoghaire,
Co. Dublin, Ireland
Telephone: (+353-1) 800245/6
Telex: 90635 TYCO EI

First edition 1983

© Copyright 1983 United Nations Environment Programme

Typeset by Printset and Design, Dublin
Printed in Ireland by Irish Elsevier Printers Ltd., Shannon

ISBN 0 907 567 16 9 Hardback
ISBN 0 907 567 17 7 Softcover

# Contents

| | | |
|---|---|---|
| Preface | | vii |
| About the Editor | | ix |
| Chapter 1 | Introduction | 1 |
| Chapter 2 | Energy and Development | 5 |
| Chapter 3 | Renewable Sources of Energy in Latin American Countries | 27 |
| Chapter 4 | Industrial Capacity and Human Resources in Latin America for producing Equipment to harness Renewable Sources of Energy | 55 |
| Chapter 5 | Marketing Renewable Sources of Energy Equipment | 63 |
| Chapter 6 | Institutional Aspects of the Use of Renewable Sources of Energy in Latin America | 77 |
| Chapter 7 | Case Studies | 87 |
| | I Small Hydroelectric Plants in Colombia | 87 |
| | II Development of Biogas Plants in Guatemala (1952-1979) | 95 |
| | III Production of Solar Collectors in Mexico | 112 |
| | IV National Alcohol Programme in Brazil | 115 |
| | V Charcoal Siderurgy in Argentina | 120 |
| | VI Las Gaviotas: An adequate technological centre that functions | 131 |
| | VII The Present State of Geothermy in El Salvador | 136 |
| | VIII Drying Coffee with Renewable Sources of Energy | 144 |
| | IX Development and Diffusion of Clay Stoves for Firewood Economy in the Peasant Area of Guatemala | 154 |
| Chapter 8 | Conclusions and Recommendations | 165 |

# Preface

IN MARCH 1977, a joint meeting took place between the Regional Offices of the United Nations Development Programme (UNDP) and the United Nations Environment Programme (UNEP) in Latin America to explore the possibilities of promoting the use of alternative sources of energy in the region. It was decided that two studies were needed: the first, on energy requirements that could be satisfied through non-conventional energy sources and the second, on the capacity for the use of existing alternative sources of energy in the region.

In March 1978, a second meeting took place in which the basic objectives of the study were agreed upon; it was decided that the study would be carried out within the framework of the UNDP project on New and Renewable Sources of Energy in Latin America (RLA/74/030) and under the technical supervision of the UNEP Regional Office. Early in 1979, the Latin American Energy Organization (OLADE) joined the project, thus establishing a joint UNDP-OLADE team. As complementary activities, among other functions, a series of surveys was carried out on Renewable Sources of Energy in nine Latin American countries. This and other information related to the subject, obtained during visits to different countries, made it possible to expand the study initially carried out by UNEP.

Several authors contributed to this study; their work is contained in the first six chapters of this book:

**Vicente Sanchez,** Research Fellow, El Colegio de Mexico, Mexico (former director of UNEP Regional Office for Latin America);

**Hector Sejenovich,** UNEP, Mexico;

**Ivan Chambouleyron,** Professor, Universidad de Campinhas, Brazil;

**Claudio Huepe,** Researcher, CENDES, Venezuela;

**Isais Macedo de Carvallho,** Professor, Universidad de Campinhas, Brazil;

**Fernando Ortiz Monasterio,** Professor, Universidad Autonoma Metropolitana, Xochimilco, Mexico;

**Luis Saravia,** Professor, Universidad de Salta, Uruguay;

**Mabel Tamborenea,** Researcher, UNICEF, Mexico;

**Jacinto Viqueira,** Chairman, Mechanical Engineering Department, UNAM, Mexico.

The editor of this book contributed to and acted as co-ordinator of the study. Contributors to the case studies contained in Chapter 7 are:

**Mercy Blanco de Monton,** Instituto Colombiano de Energia Electrica;

**Mario David Penagos,** OLADE Consultant;

**Fernando Ortiz Monasterio,** Professor, Universidad Autonoma Metropolitana Xochimilco, Mexico;

**Secretaria da Technologia Industrial,** Ministerio do Industria e Comercio do Brazil;

**Miguel A. Trosero,** OLADE Consultant;
**Jose Miguel Vellosa,** Journalist, Ceres, FAO
**Gustavo Cuellar,** Co-ordinator, Geothermy Project of OLADE;
**Alberto Chiquillo Alas,** Superintendent of Investigation, Hydroelectric Commission for Rio Lempa, El Salvador;
**Roberto Caceres,** Director, Centro Mesoamericano de Technologia Apropiada (CEMAT), Guatemala.

It should be noted that the points of view presented in this study are those of the authors and do not necessarily represent those of the Latin American Energy Organization, UNDP and UNEP. It should also be noted that the presentation of the study's results in this book is different from that in the original report presented at the First Latin American Conference on New and Renewable Sources of Energy, convened in Rio de Janeiro in 1979. This was necessary in order to bring into focus several issues of public concern. Therefore, the editor of this volume takes full responsibility for the material presented.

The publication of this study comes after the convening of the United Nations Conference on New and Renewable Sources of Energy (Nairobi, 10-21 August 1981) and the adoption of the Nairobi Programme of Action. The latter identified several areas for action, which can be grouped into five broad categories: (1) energy assessment and planning; (2) information flows; (3) research, development and demonstration; (4) transfer, adaption and application of mature technologies; and (5) education and training. The Programme of Action also recommends that the international community and the United Nations system, in particular, should develop and implement programmes and projects in these areas.

This book is for many reasons a logical follow-up to the five priorities indicated by the Nairobi Conference. In fact, the study assesses the potential of new and renewable energy sources in the Latin American region; provides and promotes information; discusses the research, developments and application of these sources of energy in the region; analyzes the desirability and difficulties of technology transfer; and provides ideas for education, institutional arrangements and training. I hope this study gives an adequate assessment of the potential of new and renewable sources of energy in Latin America and that the scientific community will accelerate the research and development efforts to meet a part of the energy requirements in the region.

Many scientists from Latin America and the Caribbean contributed in one way or another to this study; Ms Mildred Hopper from the University of Texas at Houston typed efficiently many drafts as well as the final manuscript. To all of them I would like to express my gratitude.

*Geneva, July 1982* **Francisco Szekely**

# About the Editor

Dr Francisco Szekely is a Mexican scientist, currently Assistant Professor of Environmental Sciences at the University of Texas School of Public Health at Houston. He has been Professor at the National Autonomous University of Mexico and Visiting Professor at Washington University in the United States. He collaborated for a number of years as Advisor to the Latin American Regional Office of the United Nations Environment Programme (UNEP). Since 1980, he has been Senior Advisor to the Executive Secretary of the Latin American Energy Organization (OLADE) on energy-environment issues. He has also been a consultant for UNESCO, UNDP and other international agencies, as well as several Latin American countries, particularly Brazil and Cuba. On leave from the University of Texas, Dr Szekely is working with UNEP as Deputy Director of the Regional Seas Programme in Geneva. Author of the book *The Environment in Latin America and Mexico,* he has published more than thirty books concerning the environment in developing countries.

# CHAPTER 1

# Introduction

THE SOURCES OF ENERGY that are presently termed non-conventional (or renewable) were the sole energy sources from the time *Homo sapiens* appeared on Earth until the first industrial revolution. Coal, introduced around the middle of the 18th century and incorporated into the industrial processes of the times, replaced wood and charcoal, whose intensive use had seriously deforested Europe. From that time on, people began to use fossil fuels, which were considered abundant and cheap, and consequently generated a technology based on their use and an energy-development model that has prevailed to the present.

Over the past few decades, however, increasing attention has been given to the possible use of alternative sources of energy. In the developed countries, studies have intensified on the use of nuclear fission and fusion energy, magnetohydrodynamics and renewable sources of energy (for example, solar, wind, geothermal, tidal and biomass).

Due to the effects of the 1973 'energy crisis', interest in renewable sources of energy in the developing countries has grown, especially in decentralized systems and their applications. At present, the need to study the potential of these energy sources in Latin American countries has become necessary in view of the substantial increase in oil prices; the danger of exhausting oil reserves over the next few decades; the fact that only six countries of the Latin American region are self-sufficient in oil production; the fact that the style of development adopted in the region has deprived a large majority of the population from sharing its benefits, since their levels of energy consumption place them below minimum levels for satisfying basic needs; and the fact that the prevailing model of development, in addition to making existing economic and social problems more acute, has given rise to other problems, such as deterioration of the environment due to the exhaustion of certain resources and to the unleashing of pollution processes that have proved difficult to control.

Although at first glance it might appear that all countries in the world are facing the same energy problem, more detailed analysis demonstrates that this is not the case. The developed countries are facing the problem of how to maintain consumption of high levels of energy; their concern is how to replace oil with another source of energy. However, they do not question their style of development or the ways in which such energy is consumed, despite the fact that in some countries sectorial policies exist for the conservation of such resources. In Latin America, the fundamental problem is how to provide for the energy needs of millions of people who are deprived of adequate resources. Efforts should thus be directed not towards achieving the levels of over-consumption seen in the developed countries (which would be impossible in view of the world energy situation), but rather towards obtaining consumption levels for all the inhabitants of the Latin American region that will bring about a marked improvement in the quality of life.

The solutions to the problem include the development and application of appropriate technologies for the use of conventional and renewable sources of energy and the immediate implementation of conservation measures. Neither of these paths alone can solve the problem. To a certain extent, the techniques employed for the conservation or rational management of energy will create a transitional period in which renewable sources of energy will partially replace conventional sources for certain uses and will satisfy the energy needs in situations not presently covered by commercial energy distribution networks. Thus, it will be possible to extend the period of time required to perfect equipment for using new sources and for carrying out research, so that the advanced or traditional technologies implemented will in each case constitute the best response to the energy problem at hand.

These considerations will not only introduce changes in the present style of development that will place satisfying the population's basic needs at the centre of the development process, but they will also make it indispensable to study the possibilities of developing and using renewable sources of energy in Latin American countries, to help them achieve their goals. This study is a contribution in that direction, one that focuses on the sources of energy that may be immediately applicable, in view of the relative degree of Latin American development and the progress achieved in this field on a worldwide scale. Consequently, this study is limited to the assessment of the region's potential for using the following sources of energy: direct solar, wind, plant material, biogas and small waterfalls and water flows. In view of the impossibility of studying all Latin American countries, the following were selected: Argentina, Brazil, Chile, Colombia, Costa Rica, Guatemala, Mexico, Peru, Trinidad and Tobago. They were considered to be representative of the region's efforts which are being realized in this field.

Because of the time limitations in which the study had to be performed, only partial information can be provided. It may be that significant programmes in the selected countries have not been detected or properly evaluated; nevertheless, the information compiled indicates the extent of progress made to date in research and development of non-conventional sources of energy in the region. This information was compiled from public and private universities, research institutions, energy and planning ministries, organizations responsible for national and regional science and technology programmes, business and industrial chambers of commerce, equipment manufacturers and private consultants.

The study is concerned essentially with the rural sector, since a reasonable increase in the availability of energy could play an important role in improving the quality of life of rural inhabitants and in affecting agricultural productivity. It is in this sector that the greatest deficit of energy is found in Latin America and it is here that decentralized energy systems could be widely used to complement conventional sources of energy.

# CHAPTER 2

# Energy and Development

How MUCH and what forms of energy do human beings require for their survival and development? It has been assumed that greater consumption of energy leads to improvements in well-being and development. Nevertheless, the relationship between energy consumption and development is much more complex and is mainly controlled by the characteristics and the style of development adopted.

## HISTORICAL ASPECTS

HISTORICAL variations in consumption of energy have been essentially exponential. This is particularly true in industrialized countries (Cook, 1971). Significant increases in the consumption of energy occurred when human beings began to use fossil fuels — coal in industrial times and oil in technological times. As in the case of all consumer goods, including energy, abundance and low cost led to extensive use and frequently to overconsumption. The modern industrial societies to which technological man belongs are characterized by intensive use and eventual squandering of energy. In such societies the degree of development, the quality of life and the deterioration of the environment, *inter alia,* may be measured in terms of energy consumption. A modern industrial society resembles a complex machine that transforms great quantities of high quality energy into heat and work. The efficiency with which one form of energy may be transformed into other useful forms depends essentially on the nature of the initial and final energies involved in the process and on the method employed to achieve such transformation. The efficiency of energy conversion thus varies with the nature of the process. For example, the conversion of chemical energy, such as fossil fuels and plant matter, into electrical energy is relatively inefficient (some 35 per cent at most), whereas its transformation into mechanical energy for transportation is even more inefficient, (in the order of 25 per cent). Overall conversion efficiency from primary forms of energy to the end uses in an industrial society is approximately 50 per cent.

This low degree of energy conversion efficiency partly derives from the fact that industrial societies are established on the basis of mass consumption of fossil fuels. Until the mid-18th century, energy inputs were fuelwood, agricultural residues, solar energy, wind energy, waterfall energy, the strength of draught animals and—to a lesser degree—coal, which found privileged application in the steam engine, thereby making the first industrial revolution possible. Coal thereafter prevailed as the principal source of energy until oil began to replace it at the beginning of the present century. A change in energy technology occurred with each energy transition, from fuelwood to coal, from coal to oil and, more recently, from oil to nuclear energy.

Essentially, the consumption of industrial societies is at the expense of the 'capital' composed of the earth's energy and not on a basis of permanent or non-exhaustible resources. Thus, the so-called energy crisis has emerged, in reality a crisis of fossil fuels. Bearing in mind that the developed countries concentrate more than 90 per cent of their research capacity on almost all the fields of technology, one can see that the technologies resulting from such research will be influenced by the rationales prevailing in those countries, which usually do not respond to the needs of the undeveloped countries. International division of labour, technological dependence and the application of economies of scale have reinforced this process. A change in energy technology came about only when the problem assumed the dimensions of a crisis—the abandonment of firewood for coal, the change from coal to oil and, more recently, the use of nuclear energy to solve the depletion of oil reserves.

This economic energy system, whose premise was a linear and uninterrupted development towards the kind of society prevailing in industrialized countries, was imitated by Latin American countries. This system, in many instances, did not require making the technological changes determined in other contexts and in accordance with different energy requirements. In past decades, the infinite availability of cheap oil was the energy premise on which this model was based, as it was in the societies that were being imitated.

In a sense, an economic growth path similar to that adopted by industrialized countries, characterized by enormous consumption and squandering of energy, prevented the use of other energy alternatives. This is due mainly to the fact that energy technologies have been dominated by direct application of the same rules of market economy, which maximize short-term investment. Thus, a number of costs were excluded. Evaluation of a technology should include social and political implications. The energy market itself is incapable of questioning the wisdom of selecting a particular technology and its evolution. On the contrary, it promotes the technology as long as the negative consequences of the choice can be tolerated or controlled. Consequently technologies are expanded above and beyond their socially acceptable limits; to counteract their negative effects, the market generates new technologies which, far from questioning the initial evolution, only seek to justify it further.

Beyond a certain threshold of social unacceptability, these negative effects are individually or collectively internalized by degrees. An illustration of this is the problem of pollution. Any attempt to diminish environmental pollution caused by traditional power stations without modifying the transformation processes involved or questioning the technology being used only leads to irrational management of energy resources, since energy is consumed firstly to satisfy the needs of industry and secondly to combat pollution. The role of non-conventional sources of energy should be viewed in this perspective. No reasons exist, other than those already indicated, for not having made significant efforts on a worldwide scale to consider alternatives to the conventional sources of energy.

Technologies for the use of other energy alternatives have evolved as scientific knowledge has increased. It is thus a matter of applying this knowledge through the use of the materials and methods most appropriate for each situation. The more knowledge we have of the multiple possibilities offered by available resources, the more we will be able to apply the most appropriate solution to each energy problem that arises. Appropriate technology in the field of energy is, therefore, that technology — modern, sophisticated, scientific or simple — which is capable of providing the best answer to each problem through the rational use of existing resources.

Medium- and long-term socio-economic considerations play an important role in the indeferrable need for the countries of the Latin American region to turn to technological diversity as a means of solving their energy problems. Analysis of the situation shows the great degree of technological dependence in every aspect of Latin American countries on developed nations. It also points to the overriding need for existing research organizations and institutions, which were established in imitation of the models prevailing in the developed countries, to devote their efforts to research and experimentation in appropriate technologies. The degree of importance attained by many of these institutions appears to indicate that they are in a position to do so: the lack of appropriate organizations and trained technicians should not be a serious deterrent to striving for a greater degree of independent technological development. The true limitations are to be found elsewhere — in international relations, in the model of development adopted and in the difficulty of formulating technological policies congruent with the specific conditions and needs of Latin America.

## ENERGY AND THE SATISFACTION OF BASIC NEEDS

AS A RESULT of the style of development prevailing in Latin America, energy has been used largely to satisfy the requirements of industry and the domestic needs of the middle and upper classes of society. The lower and most impoverished classes in Latin America have been compelled to satisfy their basic needs at bare subsistence levels, both in urban and rural areas. Thus, in the same manner as they build their homes from scrap materials, they use non-commercial forms of energy for cooking and heating and depend on human energy for increasing their productivity. This is particularly true in rural areas, where one of the principal sources of energy is fuelwood. And this has led to the deterioration of much of the region's forest resources.

The Brookhaven National Laboratory uses an index called the physical quality of life to compare development and energy consumption; it combines three parameters of human welfare: infant mortality rate, life expectancy and the percentage of illiteracy. A value of 100 has been assigned to the country of highest weighted index and a value of one to the lowest; the remaining countries are distributed between these two extremes. All the industrialized countries have indices ranging from 95 to 100. The disparity of the values for Latin America indicates the different degrees of development of the region's countries as measured by this parameter. Figure 2.1 illustrates the relationship between the index of physical quality of life and per capita energy consumption in 80 developing countries. The curve originates at a level of energy consumption considered to be the minimum for survival. In many countries, this consumption is derived from non-commercial sources of energy. The curve shows that an increase in the level of energy consumption has a marked effect on the quality of life up to a certain level, after which greater consumption of energy does not result in further increases in the index of physical quality of life. This energy value may be taken as that corresponding to the satisfaction of basic needs. Below it are subsistence levels and above it, the comfort levels associated with high consumption. The curve shows that one-third of the consumption corresponding to the level of satisfaction of basic needs is, or can be, covered by non-commercial sources of energy; commercial sources have to provide for the remaining two-thirds.

Available data (CEPAL, 1977) indicate that the average quality of life in several countries of Latin America (Argentina, Chile, Mexico and Colombia) is of acceptable level. Nevertheless, the disparity in the distribution of income in these countries shows that a large portion of the population — especially in rural areas — is far below energy subsistence levels. An increase in per capita energy consumption in such areas will contribute to substantial improvements in the quality of life.

The high birth rate and the accelerated urbanization process characteristic of the Latin American region have not been accompanied by an increase in food production. Agricultural production increased between 1961 and 1965 at a markedly lower rate than other economic activities. This situation has been aggravated by population growth. In 1972, Latin America, which had been self-sufficient in food production, could not satisfy its essential food needs. A similar deficit occurred in 1973. At present, statistics show that in many countries of the region a substantial portion of the population is undernourished. The main reason for the food deficit in the region does not appear to be the exhaustion of its physical sources, but rather its inability to mobilize available resources and to exact maximum yield from them. Higher gasoline and diesel oil prices have had a direct effect on the use of tractors for farming, processing of farm products, pumping of water for irrigation and on transportation of products. In addition, the increase in the cost of fertilizers manufactured from fossil fuels has affected farming in many areas, accelerating the scarcity of foodstuffs.

**Figure 2.1: Index of the Physical Quality of Life in accordance with Annual per capita Consumption of Energy in Developing Countries**

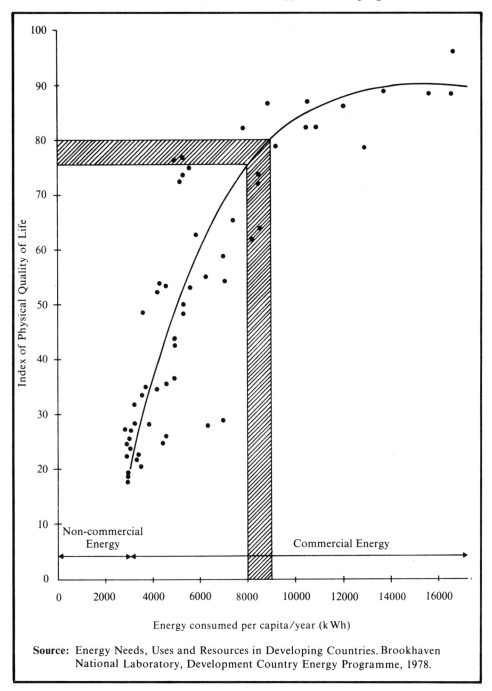

Source: Energy Needs, Uses and Resources in Developing Countries. Brookhaven National Laboratory, Development Country Energy Programme, 1978.

## ENERGY AND THE RURAL ENVIRONMENT

NATURE PROVIDES a flow of energy of varying degrees of complexity from one climatic region to another. For example, in a given ecosystem solar energy is captured by plants and transformed into chemical energy which, when consumed by animals or human beings, is transformed into energy that renders life possible. The ecosystem manages to maintain a balance by means of complex mechanisms. Thus, nature provides certain forms of energy that may be used without interfering with normal cycles of evolution and with the same speed and proportion at which they are provided. However, shortsightedness has frequently prevented respect for the long periods of time required by nature to fulfill its cycles and has led to depletion and/or deterioration of certain resources. An example of this is the use of forests and soils in the agricultural sector that has caused erosion and wash outs, which in turn have been responsible for a reduction in productive activities in other regions. Furthermore, the use of forest resources for fuelwood leads to the deterioration of a resource that could otherwise be used more rationally by replacing such use with the residues produced by industrial wood-processing.

In summary, because of the impositions of the predominant style of development — which included maximization of the society-nature relationship in the short-term and the discovery and use of increasingly complex tools for development — imbalances produced in ecosystems were of such magnitude that their own homeostatic mechanisms were not able to counteract them. In many instances, this implied great losses of energy and, in others, the need to provide the ecosystems with supplementary energy, usually in the form of fertilizers, in order to obtain the yields that the productive system required for efficient functioning.

The problems of the Latin American rural environment are not exclusively associated with energy. The relative availability of energy to satisfy basic needs (whether subsistence or production needs) is linked to the overall structure of the agricultural sector. A relative abundance of different forms of energy cannot fulfil its role of transformer if it is not accompanied by other conditions of a political and social nature. From the standpoint of energy and the technology of its use and generation, rural needs may be identified on at least two levels. The first refers to the energy needs associated with the subsistence or reproduction of the labour force; these are measured directly by the quality of life of the *campesinos* and the satisfaction of their basic needs, such as food and housing. The second level includes energy needs for production, which may be situated either in the exploitation unit itself or in the transformation of the product in question to adapt it to consumption or marketing needs.

The following list of rural needs — by no means complete — is provided to determine the extent to which non-conventional sources of energy may be able to supply such needs.

(1) Energy requirements related to the welfare or quality of life of *campesinos:*
    (a) preparation and conservation of foodstuffs;

(b) availability of water;
(c) housing, lighting and heating;
(d) communications and transportation;
(e) others.

(2) Energy requirements related to agricultural production processes:
  (a) production within the agricultural exploitation unit:
    • preparation of land: felling of trees, clearing, ploughing, levelling and the use of fertilizers and pesticides;
    • sowing and irrigation;
    • maintenance and care of animals;
    • protection of cultivated fields or animals against any agents that might deteriorate or reduce their production potential;
    • harvesting or obtaining of livestock products, such as milk, meat and eggs.
  (b) transformation of the product as raw material for another area of production or for direct consumption. The amount of energy and primary energy required essentially depends on:
    • the level of economic and technological development attained by the agricultural and livestock sector;
    • the nature of the transformation process involved (for example, the drying of cereals has an energy and technological input different from that of the canning of vegetables or the manufacture of flour);
    • demand of other production sectors or of the consumer sector.
  (c) transport and marketing of agricultural and livestock products. Transportation energy inputs depend essentially on:
    • the location of the production unit and its distance from consumption, distribution or processing centres;
    • the communications infrastructure, road network, railroads and river and maritime transportation;
    • the geographical characteristics of the region;
    • the degree of product processing and durability of the product.

Applicability of non-conventional sources of energy to the satisfaction of rural energy needs in part depends on the particular conditions, the particular application involved, the nature of the problem and, lastly, the general level of the technical culture of the *campesino* group concerned. Technical comprehension of the energy convertors being used and the possibilities of repairing and maintaining them locally are factors that must be borne in mind. In isolated areas, frequent maintenance cannot be guaranteed locally and may prove to be a limitation to the use of equipment that is otherwise perfectly adapted to the nature of the problem.

Generally speaking, technologically simple equipment (such as solar heaters, distillers and dryers) is capable of being used in any situation or location. Solar electricity and the manufacture of alcohol from plant material are examples of the most sophisticated technology as regards their production; however, this is not true of their use. Anaerobic digestors and windmills are examples of intermediate complexity as far as their technology is concerned.

# STYLE OF DEVELOPMENT IN LATIN AMERICA

## Basic Characteristics of the Predominant Style of Development

Since the early 1950s, the developing countries have increased their efforts to overcome the chronic problems of their economies. They embarked upon new production methods as a means of achieving the much-sought goal of industrial development attained by the developed nations, thereby establishing the present-day style of development.*

The predominant style of development in Latin America over the past thirty years has essentially been an imitation of the style adopted in Europe and the United States, both in general terms and with specific reference to technology and energy. This style is based on the promotion of industrialization to attain the rapid growth required to bring about the desired well-being of the population. Consequently, because it failed to take into account the particular social and economic characteristics of the Latin American region, the adopted style of development did not produce the hoped-for regional effects and had varying repercussions in each of the countries involved.

The principal premises upon which this style was based were founded on the supposition that significant incentives to the industrial sector through the use of modern technology would bring about a highly dynamic increase in all productive activities that would, in turn, propagate technological progress throughout the economy. Such a modernization process would raise the qualifications of both labour and management and thereby lead to new and highly productive investment. The expected reduction of costs would place a greater quantity of goods within reach of the population and consequently increase its living standards. In turn, the incorporation of advanced technology would permit future generations of domestic technology, which would promote self-sustained development. Certain reforms in the agrarian structure, combined with the incorporation of capital and technology, would end stagnation in the agricultural sector and convert it into a consumer of industrial products.

The events of the past thirty years show that despite significant advances in the growth of productive activities, the well-being of the population was not improved. The system as a whole proved to be sufficiently flexible to absorb the substantial changes deriving from the influence of the world economy without suffering serious structural damages; it also showed itself to be receptive to the

---

*From a predominantly economic standpoint (which is limiting but useful for the purposes of the present study), a style of development may be defined as the manner in which within a given system, human and material resources are organized and assigned with the objective of resolving the question of what, for whom and how to produce goods and services. This definition takes into account: (1) the structural basis of production (that is, the sectoral structure of the product and employment, the various technological strata and the predominant type of external relations) and (2) the system's dynamic elements, which are drawn from analysis of the level and composition of demand and from the level and distribution of income. These elements are structurally closely interlinked (Pinto, 1976).

incorporation and development of complex technologies. However, it proved to be inefficient in its ability to pass on the fruits of these advances to all sectors of the population. This fact is related to the comments made in the definition of the style of development with reference to 'for whom, what and how' to produce. In a market economy, such as that of Latin America, consideration of 'for whom' is intimately related to the amount and distribution of income and to the amount and composition of demand.

There is no doubt that the industrialization process fulfilled the demands of the highest income groups—the minority groups that are most represented in the market. For example, in 1960 these groups represented 30 per cent of the higher income population and received 72.5 per cent of total income, whereas 50 per cent of the lower income population received only 13.4 per cent of the total income (CEPAL, 1977). From 1960 to 1975, this situation worsened since the top income-receiving group in Latin America (50 per cent of the population) increased in participation in 8 per cent of the total income, whereas the 90 per cent bottom of the population lost more than 2.5 per cent from their participation; the intermediate stratum suffered a loss greater than 6 per cent of the total income generated by the region (CEPAL, 1979). Futhermore, as the minority groups wield a substantial portion of political and economic power and also serve as a kind of 'example' (insofar as the adoption of cultural habits, scale of values and consumer patterns of the upper and middle classes of the developed countries are concerned), they have influenced the development of a modern sector of the economy generally linked in various ways—financially and technologically—to the industrialized countries.

The iron and steel, petrochemicals, heavy machinery and chemical industries were set up in several countries since they were considered to ensure the greatest potentiality for an entire dynamic industrial and economic structure. At the same time, power stations were established and expanded to provide these industries with the energy they required.

One of the most important phenomena, derived from the evolution of forms of unequal competition, is the development of economies of scale. Centralization made possible substantial reductions in costs and, consequently, also brought about concentration of the supply of energy, the type and characteristics of which had to be congruent with the dominant technology. This in turn led to the predominant use of oil and gas energy as the appropriate response to spatial concentration. Although, in certain cases, attempts were made to encourage industrial deconcentration by putting into practice the theory of poles of development, these initiatives were not as successful as expected, since they were not able to mobilize already existing investment, received little publicity and, even when realized, repeated the same tendencies towards concentration that had been demonstrated in the principal centres.

Qualitative growth of production in the region was of the greatest importance. In the period from 1950 to 1979, the economy grew at an average annual rate of 5.5 per cent, a level significantly higher than the most optimistic predictions. The production of commercial energy grew at an annual cumulative rate of almost 7 per cent. Industry's dynamic growth during this period increased its share in the national product from 18 per cent to 24 per cent. Naturally, this industrial growth process was not uniform, since substantial differences may be noted among the countries in question. For example, a distinction must be made between countries with large markets based on well-integrated industrial structures and countries whose production is concentrated in a few products, with incipient industrial structures and serious difficulties in continuing to substitute imports. Nevertheless, in most countries, the new industrial structure made it possible to satisfy a large portion of the demand for consumer goods. In some industries, intermediate goods were produced and in others, even capital goods. Adaptation of the productive structure to the demands of the high-income sectors led to the manufacture of products they required, leading to situations in which the problems of relative saturation and constant introduction of new products to satisfy increasingly sophisticated demands soon replaced any idea of directing the productive structure towards satisfying the basic needs of the population.

Although it is true that several aspects of the modernization process filtered down to other, lower income sectors of the population (for example, the partial mechanization of agriculture, the establishment of small industries and the adoption of new cultural patterns, particularly in terms of health and education), a large portion of the population of the region continued to remain outside the mainstream of the process. Such results may be attributed to the fact that the style of development briefly described did not take this portion of the population into account when considering 'for whom' to produce and was consequently incapable of solving its age-old problem of poverty. The situation became even more critical as distribution of income became more unequal and huge masses of rural dwellers, unable to cope with the problems caused by the reduced size of their land parcels and increasing mechanization, became permanent and/or stationary migrants. These migrants set up precarious dwellings around the large cities in the vain hope of obtaining the employment that the predominant capital-intensive industry was unable to provide. The principal urban centres thus grew disproportionately and gave rise to concentrations that in some cases accentuated already existing geographic imbalances and created them in others.

Without taking up the question of a 'dual society'—which implies both a 'modern' and a 'traditional' sector of the economy, inasmuch as both are inseparable parts of a single and structurally interdependent system—it should nevertheless be pointed out that the rapid growth which took place in the sector described as the most modern and most dynamic was largely possible because of the transfer of capital generated in what is known as the 'traditional' sector.

In summary, despite the significant growth achieved by the adopted style of development in productive activities, these considerations demonstrate that serious difficulties still persist to prevent the fruits of such development from bringing about the general well-being of the population.

## Distribution of Income and Basic Needs

Although there are limitations in dealing with Latin American averages, a few figures are provided here that demonstrate the seriousness of the problem (CEPAL, 1976). In 1970, the poorest 50 per cent of the population had an income of only US$112 per capita, whereas the wealthiest 5 per cent had a per capita income of $2,600, almost equivalent to that of the developed nations. This disparity is all the more notable in view of the fact that 20 per cent of the poorest population in 1970 received only 2.5 per cent of the income, with an annual average of $55 per capita.

However, the seriousness of the matter lies not only in the existing unequal distribution, but also in the inflexibility of the structure itself which makes it incapable of change. Thus, in the period from 1960 to 1970 in the countries where CEPAL studied income distribution (that is, those constituting the majority of the region's population — Argentina, Brazil, Colombia, Chile, Honduras, Mexico, Paraguay and Venezuela), per capita income rose from $345 to $440. As a whole, income increased by $25,406 millions. Of this total increase, 20 per cent of the low-income population received only 0.4 per cent. Fifty per cent of the poorest population received 15.8 per cent of the total increase, whereas the wealthiest 15 per cent of the population received 47.8 per cent of the total. From another standpoint, such distribution of income confirms the specific orientation of the adopted productive structure toward supplying the demands of a small, high-income sector.

As far as satisfying the needs of health, housing, education and nutrition, one can observe that each of the countries in question shows differences between these sectors. However, available data demonstrate that although significant progress has been achieved, it continues to be unsatisfactory. For example, in the field of education, the efforts made by the majority of governments over the past thirty years have been considerable but the results achieved have been meagre indeed. The situation is similar for health and housing. Even though overall indicators per country or for the entire region point to improvement, the problem of regional inequalities in each country persists as a result of concentrating most efforts in the principal urban centres (CEPAL, 1975). With regard to both calories and protein consumption, the Latin American population, on average, barely exists above subsistence level. Nevertheless, it may be noted that the poorest 50 per cent of the population does not even arrive at an acceptable minimum.

From a cultural standpoint one more important piece of information should be considered in dealing with this problem. Industrial growth, and particularly structural change of the sector, requires the encouragement of consumption as a means of increasing the market for the sale of products. The 'example' effect referred to previously also functions in the lower income stratum, so that the lack of satisfaction of basic needs is compounded by a subjectively experienced greater lack of satisfaction created by the discrepancy between consumer aspirations (which are intensified by the communications media) and the realities of subsistence levels of income.

## Technology and Employment

The technical changes introduced into the system's productive processes are in a close relationship of interdependence with the general functioning of the system itself and thus may be both cause and result of social, political and economic changes. The differential structure of income and its inflexibility with respect to change are partially derived from the characteristics of the technologies evolving from the style of development selected by the developing countries and are oriented towards economic growth. These technologies are capital-intensive and at times highly sophisticated. They are appropriate for the most developed countries since, generally speaking, they do not require extensive use of manpower and thereby convert the industrial sector into a productive sector—which, in combination with the situation described in the rural sector, has given rise to serious difficulties in providing productive work for the burgeoning, economically active population.

Insofar as employment is concerned, unemployment figures do not appear to indicate too serious a situation. However, such is not the case with partial unemployment, including those sectors whose productivity is infinitesimal. Total and partial unemployment together come to 35 per cent of the agricultural labour force and to 28.4 per cent of the remaining labour force. Estimates made by CEPAL show the composition of employment and product by technological categories at the end of 1960. These figures reveal the structural heterogeneity existing in the productive sectors. For Latin America as a whole, the division of technologies into modern, intermediate and primitive shows that the modern sector of the economy of Latin America employs 12.4 per cent of the labour force, which produces 53.5 per cent of the total product. It would thus appear that the great technological development of recent years has provided only 12.4 per cent of all employment. In agriculture, this category of technology provided 6.8 per cent of employment, in manufacturing, 17.5 per cent and in mining, 38 per cent.

The percentages corresponding to the share of this category in total production reveal its decisive significance, despite the small amount of labour employed. In manufacturing, it is responsible for 62.5 per cent of the total product; in agriculture, 47.5 per cent, nevertheless accounting for almost all of the most important export products; and in mining, 91.5 per cent. The modern technology sector consists of a small number of corporations that control a large portion of the market and consume the largest amounts of energy.

## The Agrarian Sector

In the field of agriculture, the consequences of the predominant style of development have been even more serious than in the industrial sector. It was imagined that the predominance of industrial activities as a strategy for accelerating development would be followed by corresponding modernization and increased accessibility of the lower income sectors to ownership of the farmland of which they had been deprived by the rigid ownership patterns characteristic of the early 1950s. Generally speaking, in the period studied, the agricultural sector increased its activities less than other fields. Consequently, it might be assumed that natural resources were subject to little pressure whereas, in reality, heavy pressure was exerted upon them which substantially diminished their productive potential; deterioration of forests, loss of soil fertility, pollution of rivers and other forms of environmental degradation were evident.

Since a market economy was also predominant in this sector, land use was not planned and the desire to obtain maximum yields as soon as possible was the guiding criterion, thus making it possible for two apparently contradictory situations to exist simultaneously. On the one hand, certain resources were not sufficiently exploited and on the other, more important resources were over-exploited, degraded and squandered. This has brought about a sharp reduction in the productive capacity of certain natural resources in Latin America and has also led to a loss of productive opportunities.

With such a state of affairs, one can do little more than question the relationship that this system as a whole has established with nature, inasmuch as the desired results are not being obtained. Accordingly, particularly during the 1960s, changes in agrarian structures were carried out in some countries by means of reform programmes that achieved varying degrees of success.

At the same time, the world market began to impose changes in the kinds of agricultural production being carried out. Thus, the production of industrial crops used as raw materials by agro-industries was intensified, as was the production of export crops requiring processing by means of appropriate technologies. Management of agro-industries is usually in the hands of large agricultural enterprises that use oligopolistic marketing structures to lower the prices of the agricultural products they require as raw materials. These crops are usually produced by small-scale producers or tenant farmers. The decline in prices brought about by agro-industrial firms and by official price policies designed to support industrialization made it impossible for the majority of the rural masses in the region to obtain minimum subsistence levels from the sales of their products. This situation had two important consequences: mass migration of rural dwellers to the cities and extensions of the agricultural frontier to land that in many cases had not been used to advantage previously, but rather used in accordance with the market requirements of the moment. Thus, the model of rapid industrial growth intensified Latin America's already unbalanced development, not only with regard to individual countries but also with regard to the various social strata that make up the region's societies.

## THE ENERGY SITUATION IN LATIN AMERICA

ONE OF THE PRINCIPAL energy sources used throughout the world to-day is *oil*. Nevertheless, CEPAL (1975) reports that "... in absolute figures, oil consumption in Latin America is rather meagre, since this area consumed only 6.6 per cent of world consumption of oil in 1965 and 6.3 per cent in 1980, which is approximately in the same proportion as the increase in world oil consumption between 1950 and 1965. Such small percentages reflect Latin America's relatively small participation in world income, the comparatively high proportion of non-industrial activities in its economy, the widespread use of production techniques requiring comparatively little energy and the relatively intensive use of vegetable fuels in the region— approximately 20 per cent in 1972."

Table 2.1 shows the world energy situation for 1970, with a projection to the year 2000. In 1970, Latin America consumed energy equivalent to 254 million tons of coal equivalent—3.7 per cent of the world total. However, this figure is expected to increase sevenfold by the year 2000. The question remains of how the region is going to fulfill its increasing demand. However, if this is the comparative situation on the world level, the reality on a regional level and with respect to the source of production is another matter.

In 1979, total energy consumption in Latin America was equivalent to 319 million tons of oil equivalent. Distribution by source is given in Table 2.2. It is clear that Latin American dependency on oil (66 per cent of total) is high.

To evaluate the energy situation in Latin America, a valid question would be: does the region produce enough oil to supply its own internal demand? The answer to this is not easy. Regional production of oil in 1978 amounted to 1,796 million barrels per day—a figure higher than that of consumption (Table 2.3). However, if the countries that were oil exporters at that time are excluded (Venezuela, Ecuador, Bolivia, Trinidad and Tobago, Mexico, and Peru), consumption (1,327.8 million barrels) will be seen to be greater than production for all the other countries of the region. In 1973, the 19 countries deficient in hydrocarbons imported 42 per cent of the oil they consumed. In terms of its energy balance sheet, the Latin American region uses the greatest amount of hydrocarbons as far as percentages are concerned, relative to the rest of the world.

Mention should be made at this point of the market imbalances existing in Latin America between the commercial energy consumption structure and the capacity of local sources to satisfy such consumption. In the case of oil, of the 28 countries listed in Table 2.3, six are self-sufficient, another five are partially self-sufficient and the remainder import all their oil supplies.

*Water resources* are also used as a source of energy in Latin America. This resource exists in most of these countries in such abundance that Latin America could be identified as a region with hydroelectric energy potential. However, hydroelectricity is still little used and is especially concentrated in large hydroelectric power stations rather than in the use of small water flows, which are abundant and well-distributed geographically. So far, this resource has essentially

Table 2.1 World energy situation in 1970 and 30 years later
(Consumption of energy in millions of tons of coal equivalent)

| Location | 1970 | | | 2000 | | |
|---|---|---|---|---|---|---|
| | Consumption of Energy | Percentage of Market | kW per Capita | Consumption of Energy | Percentage of Market | kW per Capita |
| Africa | 109 | 1.6 | 0.29 | 564 | 1.6 | 0.63 |
| Asia | 1,043 | 15.3 | 0.46 | 13,378 | 37.8 | 3.23 |
| Central America and Mexico | 142 | 2.1 | 1.94 | 1,216 | 3.4 | 6.17 |
| North America | 2,472 | 36.3 | 9.90 | 7,934 | 22.4 | 21.76 |
| South America | 112 | 1.6 | 0.47 | 710 | 2.0 | 1.37 |
| Western Europe | 1,350 | 19.8 | 3.44 | 4,367 | 12.4 | 9.04 |
| Eastern Europe | 460 | 6.7 | 4.04 | 1,623 | 4.6 | 11.60 |
| Oceania | 78 | 1.1 | 3.67 | 364 | 1.0 | 9.44 |
| Soviet Union | 1,055 | 15.5 | 3.97 | 5,242 | 14.8 | 14.51 |
| WORLD | 6,821 | (100) | 1.71 | 35,400 | (100) | 4.96 |

Source: Economic Commission for Europe (1975) Document ECE/Env./R.31.

**Table 2.2  Energy Consumption in Latin America by Source in 1979**

| Energy Source | Million Tons Oil Equivalent | Per cent of Total |
|---|---|---|
| Oil | 211.8 | 66 |
| Natural Gas | 44.0 | 14 |
| Hydroelectricity | 46.8 | 15 |
| Coal | 16.0 | 5 |
| TOTAL | 319.4 | 100 |

Source: US House of Representatives Committee on Interstate and Foreign Commerce (1980). *The Energy Factbook.* US Government Printing Office, Washington, DC.

been used for the generation of electric power. It was estimated that, before 1973, the electric power that could be generated economically in the region was of the order of 2.8 million GW, 25 times the total generation of hydroelectric power recorded in 1973. As a consequence of the increase in cost of thermal power stations, many hydroelectric projects have now become profitable operations and it is therefore possible that the figure of 2.8 million GW will increase (CEPAL, 1975). OLADE (1981) has carried out the first inventory of hydroelectric resources in the Latin American region using an homogeneous methodology in its evaluation. The results of those estimations are included in Table 2.4.

Another resource of lesser and more localized use is *natural gas*. Large reserves of this resource (a total of 1.930 billion cubic metres in 1970) are concentrated in Venezuela (45 per cent), Mexico (12 per cent) and Argentina (9 per cent). Gas deposits also exist in Bolivia, Brazil, Colombia, Chile, Peru, and Trinidad and Tobago. Gas production is generally associated with the off production, which has often led to its being considered as a byproduct and to large-scale wastage.

Measured reserves of *coal* are estimated at 4,100 million tonnes and potential reserves at 60,000 million tonnes. Present-day production is of the order of 12 million tonnes per year. The largest deposits have been found in Argentina, Brazil, Colombia, Chile, Mexico, Peru and Venezuela.

Although it appears that *geothermal energy* exists throughout the western part of the continent, little is known of its potential. Only a few countries have undertaken exploration and/or exploitation: Mexico, El Salvador, Guatemala, Nicaragua, Panama and Chile.

Prospecting for *radioactive minerals* is of very recent date; Argentina, Brazil and Mexico are the most advanced countries in this field. As of 1980, estimated uranium reserves (uranium extractable at less than US$130 per kilogram) were of

## Table 2.3 Latin America: Production, consumption, exports and imports of oil in 1978 (in $10^6$ barrels)

| Country/Territory | Production | Consumption | Exports | Imports |
|---|---|---|---|---|
| Dutch Antilles | — | n.a. | — | 197.5 |
| Antigua | — | n.a. | — | 5.8 |
| Argentina | 162.8 | 173.5 | — | 15.3 |
| Bahamas | — | n.a. | n.a. | 182.5 |
| Barbados | 0.2 | 1.5 | 0.02 | 0.6 |
| Belize | — | n.a. | — | — |
| Bolivia | 11.8 | 7.6 | 2.86 | — |
| Brazil | 58.52 | 385.3 | — | 328.9 |
| Cayman Islands | — | n.a. | — | — |
| Colombia | 47.74 | 46.46 | — | 8.8 |
| Costa Rica | — | 6.92 | — | 58.4 |
| Cuba | 1.8 | 44.5 | — | 58.4 |
| Chile | 7.5 | 36.7 | — | 27.1 |
| Dominica | — | 4.5 | — | 5.4 |
| Ecuador | 73.6 | 30.9 | 44.79 | 0.17 |
| El Salvador | — | 4.5 | — | 5.4 |
| Grenada | — | n.a. | — | n.a. |
| Guadalupe | — | n.a. | — | — |
| Guatemala | — | 8.35 | — | 5.83 |
| Guyana | — | 4.59 | — | — |
| Haiti | — | 1.69 | — | — |
| Honduras | — | 3.79 | — | 3.4 |
| Jamaica | — | n.a. | — | 12.03 |
| St. Kitts | — | n.a. | — | — |
| St. Lucia | — | n.a. | — | — |
| Martinique | — | — | — | — |
| Mexico* | 485.29 | 402.0 | 300.0 | — |
| Montserrat | — | n.a. | — | — |
| Nicaragua | — | 6.1 | — | n.a. |
| Panama | — | 13.36 | — | 16.7 |
| Paraguay | — | 3.09 | — | 2.5 |
| Peru** | 73.0 | 45.6 | 25.0 | — |
| Puerto Rico | — | n.a. | — | 99.93 |
| Dominican Republic | — | 14.59 | — | 11.29 |
| Surinam | — | n.a. | — | — |
| Trinidad and Tobago | 83.7 | 4.94 | 49.4 | 56.8 |
| Uruguay | — | 12.85 | — | 15.06 |
| Venezuela | 790.35 | 109.5 | 482.16 | — |
| St. Vincent | — | n.a. | — | — |
| Virgin Islands (US) | — | n.a. | — | 168.0 |
| Virgin Islands (UK) | — | n.a. | — | — |
| Latin America and the Caribbean | 1,796.3 | 1,372.8 | 904.3 | 1,285.81 |

n.a. = Not Available
— = Very Small
\* = On 1 September 1980, Mexico reported 2.2 million barrels/day oil production
Source: Lopez Portillo (1980) J. Informe Presidencial, Mexico, September
\*\* = *Petroleum Economist* (1979) 'Peru' July

**Source:** The Petroleum Publishing Co. (1980) *International Petroleum Encyclopaedia*, Tulsa, OK.

### Table 2.4 Hydroelectric Resources of Latin America

| Country | Potential (MW) | Installed Capacity (MW) | Utilization (per cent) | Thermal Equivalence (MWe) |
|---|---|---|---|---|
| Argentina | 45.000 | 3.169 | 7.0 | 3.233 |
| Bolivia | 18.000 | 0.242 | 1.3 | 1.293 |
| Brazil | 213.000 | 24.137 | 11.3 | 15.302 |
| Colombia | 120.000 | 3.072 | 2.6 | 8.620 |
| Costa Rica | 8.900 | 0.400 | 4.5 | 0.639 |
| Chile | 12.000 | 1.470 | 12.3 | 0.862 |
| Equador | 22.000 | 0.214 | 1.0 | 1.580 |
| El Salvador | 0.850 | 0.250 | 28.6 | 0.061 |
| Guatemala | 9.900 | 0.105 | 1.0 | 0.711 |
| Guyana | n.a. | n.a. | n.a. | 0.862 |
| Honduras | 2.800 | 0.109 | 3.9 | 0.201 |
| Mexico | 25.250 | 5.200 | 20.6 | 1.814 |
| Nicaragua | 2.950 | 0.130 | 3.4 | 0.211 |
| Panama | 2.900 | 0.273 | 2.1 | 0.208 |
| Paraguay | 17.000 | 0.190 | 1.3 | 1.221 |
| Peru | 58.000 | 1.815 | 3.1 | 4.167 |
| Surinam | n.a. | n.a. | n.a. | 0.019 |
| Uruguay | 7.000 | 0.281 | 4.0 | 0.503 |
| Venezuela | 36.000 | 2.680 | 7.4 | 2.586 |
| Caribbean | 16.000 | 0.486 | 3.0 | 0.269 |
| LATIN AMERICA | 617.550 | 44.233 | 7.16 | 44.362 |

Source: OLADE (1981)

n.a.: Not Available

the order of 53,000 tonnes for Argentina, 73,000 for Brazil and 1,900 for Mexico. The Atucha and Rio Tercero power stations in Argentina will consume 50 and 80 tonnes of fuel, respectively, per year.

*Biomass* accounts for a significant amount of the energy used in the region (approximately 20 per cent.) Unfortunately, indiscriminate use has brought about mass destruction of a highly valuable ecological and energy resource: the forest land of Latin America. The potential of biomass in Latin America is large indeed. Brazil has begun to profit from these resources with its alcohol programme. However, a total account of the size of these resources is difficult.

# THE OIL CRISIS AND ITS IMPACT ON LATIN AMERICA

THE SO-CALLED ENERGY CRISIS has affected the countries of Latin America in different ways and with varying intensity. Firstly, the increase in oil prices has affected the prices of other energy sources. This phenomenon, which scarcely affects the higher income strata, may become a serious matter to low-income groups who must restrict their use of all kinds of fuels. The characteristic imbalance in the region in the consumption of energy is further increasing the gap between social sectors and endangering the development process itself. Secondly, any increase in the price of oil (and any other form of energy that local industry uses) is reflected in the prices of manufactured goods, which then rise beyond the reach of the low-income groups.

In rural areas, this situation encourages the consumption of fuels such as firewood and agricultural wastes. The former implies even greater pressure on an already threatened resource and the second prevents manure and other wastes from being recycled into the earth. If to this is added the greater cost of industrial fertilizers, the result is less agricultural productivity which, in turn, has repercussions on the already deficient nutritional levels of the population.

The style of development that has been selected is based on the premise of an almost limitless availability of oil at low prices. World oil reserves are a topic of unending discussion and various current estimates place them at an average of 650 billion barrels. All estimates indicate that world crude-oil production will reach its maximum point before the year 2000. If to these considerations are added that (a) the availability of oil will have greater strategic value in the future than at present, (b) its price will be that much higher as reserves dwindle and (c) the surpluses of the oil-producing countries almost certainly will be destined for the industrialized nations rather than for the countries of the region, then the prospects for all Latin American nations are dismal indeed. With the exception of a few countries, the situation is alarming and it is unlikely that notable improvement will take place over the coming decades in view of the turn that the persistent energy problem is taking.

As far as current consumption is concerned, it is known that a single inhabitant of the United States consumes between 10 and 20 times the amount of commercial energy consumed by the average inhabitant of Latin America; Europeans consume between 5 and 10 times the amount of their Latin American counterparts. If one bears in mind that the energy consumed in the United States and in Europe is almost exclusively of fossil origin—more precisely, oil—the impossibility of development with the system described appears in all its cold reality. These figures also show that if the final goal is the region's progress, the problem of development should be considered on the basis of alternative sources of energy. This leads to the inevitable adoption of other styles of development and reconsideration of the use of the existing technology that will make it possible to increase the use of energy resources that have not been exploited because of the abundance of oil.*

At the time of the 1973 oil crisis, the first impulse of the oil-importing countries possessing coal, natural gas and water resources was to convert their thermal power stations, run by fuel oil, to coal or natural gas and to increase the use of their water resources. In the first instance, these two energy alternatives would have cost both time and heavy investment since conversion to coal energy would have cost from US$50 to $70 per installed kilowatt. If such conversion were effected, it would solve the oil problem of the large urban centres and of interconnected systems, but not the problem of supplying energy to small population centres or isolated communities since the cost of supplying energy to them by interconnection is very high in some countries.

The idea of building large hydroelectric projects would, in general, fulfill the same objectives. Nevertheless, the disadvantage here is that hydroelectric projects require more extensive and detailed study since they must be adapted to the environmental conditions of the site on which they are constructed. They must also establish complementary uses for the water resources used and must be adapted to the characteristics of the system into which they are to be integrated. Such projects provide more alternatives to possible models of development than those previously mentioned since they are of multiple use. It is also true that in view of increasing oil prices, dams whose construction formerly appeared to be extremely expensive, now appear to be quite feasible. However, since construction times for public works projects are long and the delivery of equipment is slow in Latin America—usually because of financing problems—it is hardly likely (as in the case of conversion of thermal plants) that construction of hydroelectric plants would constitute a significant saving in oil for the remainder of the present and part of the coming decade.

Similar problems arise with respect to nuclear power plants, which are now competitive and can also be complemented with water-powered plants. Nevertheless, nuclear plants have the disadvantage of requiring extremely high initial investment, most of which must be in foreign exchange, not to mention the polluting effects they may cause to the environment. The situation of oil-importing countries that do not possess coal or natural gas is obviously much more problematic.

These considerations serve to emphasize the gravity of the situation of the Latin American region as a result of the 'oil-intensive' model followed up to the present time. They also point out the almost insuperable difficulties that the presently developing countries will experience in seeking development by following the same model as the developed nations. Constant increases in the

---

*In CEPAL's *America Latina y los Problemas Actuales de Energia*, it was mentioned that, with the measures adopted by various countries and the Organization of Petroleum Exporting Countries (OPEC) on prices and limitations on the supply of this resource, the discomfiting external dependence of several countries in the region with regard to their energy needs was brought to light. It was also noted that the oil-importing countries soon saw their balance of payments rapidly and negatively affected because of the 1973 oil crisis.

price of crude oil and its byproducts, combined with the financial situation of most of the importing countries—which will prevent them from buying oil and also from making the other investments required for their economic growth—make it imperative to seek new energy alternatives that will be compatible with the requirements and objectives postulated by the new international development strategy.

# CHAPTER 3

# Renewable Sources of Energy in Latin American Countries

THIS CHAPTER presents the results of research carried out in nine Latin American countries (Argentina, Brazil, Chile, Colombia, Costa Rica, Guatemala, Mexico, Peru, and Trinidad and Tobago) on the existing capacities in these countries with regard to technologies for the exploitation of renewable sources of energy. The investigation was aimed at determining the overall number of public and private projects being implemented or planned. Although the information gathered may be considered partial, it serves as an indication of the degree of progress attained in the study and exploitation of these sources in the countries visited.

## SOLAR ENERGY

MOST OF THE countries visited have measured the intensity or number of hours of sunlight. Generally, this information has been compiled over several decades from a considerable number of stations. Nevertheless, the available data have not been properly processed and can rarely be considered accurate or reliable. In some countries, work is being initiated on the compilation of such data which, together with other meteorological data such as cloud cover and humidity, will indirectly provide information on solar radiation. In recent years, the interest in direct measurements of solar radiation has increased and stations are being set up to determine the amounts of energy received from the sun. In all cases however, sufficient information on the availability of solar energy during the different periods of the year has been obtained to plan experiments on the exploitation of solar energy.

The exploitation of direct solar energy is the area in which the most significant progress has been made in the countries included in this study. In many of them, locally manufactured solar equipment is already being sold and many prototypes are also ready for the market. Most of the research and development taking place at present could be included in what is known as 'soft', 'appropriate' or 'intermediate' technologies, which are generally characterized by being labour-intensive, not producing harmful effects on the environment and of simple design and manufacture. Few research and development projects exist that involve advanced or capital-intensive technologies, such as production of electricity or hydrogen using solar equipment.

## Argentina

Stations measuring the number of hours of sunlight have existed throughout Argentina for decades. This information led to the plotting of a solar radiation map for the entire country. Measurements using pyrometers were also taken for brief periods of time in a small number of stations, the results of which were used in plotting the radiation map. A plan exists to install a complete measurement network and 40 electronically integrated stations are currently being set up to measure total radiation. A computerized central data bank and a calibration laboratory will also form part of this network; thirty stations with pyrometers are presently functioning. The information collected over the past years has not been processed yet and it is expected that measurement errors will be considerable.

Intensive work is being carried out in developing solar dryers for fruit, cereals and tobacco. Water heaters for domestic use are produced locally and sold throughout the country. The heating of water for industrial use has also received attention and a study is underway for its use in the extraction and separation of uranium. Experiments have been made on solar water pumping for use in extracting industries and on farms. A distiller of brackish water (100 square metres) is being constructed; this technology has already been perfected for use on a smaller scale. The study and development of passive solar systems in architecture has received particular attention over the last few years. Official interest has been expressed in this field and at present, plans are being finalized for the construction of solar houses in high, dry and cold areas and for low-cost housing in dry, temperate areas. With respect to the generation of electricity, an optical concentration system is being developed to produce steam at a temperature of 400°C to feed a thermal electricity generation system.

## Brazil

The first solar radiation map for the entire country was published in 1978 by the Meteorological Section of the National Institute for Space Activities. This map was prepared by using Ångström-type correlations and was essentially based on information concerning the number of hours of sunshine. Few stations at present are measuring solar radiation; however, a great number of stations dependent on the National Meteorological Department are measuring the number of hours of sunshine.

Solar dryers have been developed for drying grains with a capacity of up to 60 tonnes. An important programme conducted by government agencies has been initiated to use large-scale solar dryers for soybeans, beans, coffee and cocoa. The drying of fish has also been the subject of experiments on a pilot scale. Water heaters for domestic use are produced locally and are available commercially. Research is being carried out to reduce costs and utilize new construction materials. Collectors for space-heating, distillation, refrigeration and thermal equipment have been developed. Intensive work has been carried out on air heaters for industrial use and forced convection systems now exist as prototypes. No great industrial effort has been made to develop refrigeration equipment, although research in this field has been carried out in universities. With regard to solar electricity, battery-charging equipment is produced and sold locally, although with imported silicon photocell components.

## Chile

A national network of 120 stations exists for measuring the number of hours of sunshine, many of them possessing actinometers. In addition, there are 35 stations possessing data on total radiation recorded over the past three years. Radiation maps are being prepared at the Universidad Tecnica Federico de Santa Maria, which possesses information compiled since 1955 and is responsible for its processing. Five technicians are employed to carry out the mapping and conversion required to estimate daily solar radiation on the horizontal plane. The need to carry out these operations manually has caused a certain delay in data conversion.

Chile has a long tradition in the use of solar energy and studies on the heating of water for domestic and industrial use have been proceeding for many years. At present, commercial application is being initiated. Solar distillation and its associated technology have been developed and are being used to make brackish water potable. Chile has experience at the prototype level in the drying of agricultural products, particularly fruit, and the feasibility of large-scale use has been demonstrated. No significant efforts have been made with regard to refrigeration and solar electricity.

## Colombia

The Colombian Institute of Hydrology, Meteorology and Land Improvement (HIMAT) is presently in charge of the country's meteorological stations. Two hundred and seventy of these stations are equipped with Campbell Stokes Sunshine Recorders. The data collected by these stations have not been processed systematically and at present cover a period of 18,418 months. Based on this computerized information, the DER Unit of the Nuclear Affairs Institute are preparing a sunshine map for Colombia — a map indicating the variation in sunshine during different months of the year at the observation sites. This information will be combined with other meteorological data (such as those concerning temperature, relative humidity, wind and cloud cover) as a means of calculating radiation. A study has been concluded on the pattern of sunshine on the plain of Bogota prepared by the Research and Meteorological Application Section of HIMAT. This project consists of a time (ten years) and space analysis of the ranges and distribution of sunshine, taking account of monthly and daily averages recorded by all stations in the area.

The drying of grains, seeds and foodstuffs through the use of solar energy has been a topic of general interest in research and development activities in Colombia. Dryer prototypes already exist for diverse uses, although they are not yet extensively used. Distillation processes have been employed on an scale for many years. Distillers are being developed with local technology and are expected to be put on the market shortly. The most advanced solar energy projects in Colombia are concerned with the heating of water for domestic use. Water heaters are already on the market and are installed in government-supported housing projects. Cadmium-sulphide solar cells for rural telephone communications, as well as pyrometers and periheliometers, are also under construction. However, no significant efforts have been made in the areas of refrigeration and solar electricity.

**Costa Rica**

No solar radiation map has been plotted for Costa Rica to date; however, plans are being made at the University of Costa Rica for producing one in the near future. The University's School of Electrical Engineering has been measuring solar radiation in certain areas of San José. The most important research and development work has been carried out at the National University with reference to concentrators for the cooking of foodstuffs and to water heating and drying systems.

**Guatemala**

No solar radiation maps exist for different regions. At present, the National Institute of Seismology, Vulcanology, Meteorology and Hydrology (INSUVIMEH) operates 16 stations measuring solar radiation. The information collected is now being processed. No solar equipment is yet on the market and research carried out to date has been applied essentially to hot-air drying of grains and foodstuffs, water heating for domestic use and solar ovens.

**Mexico**

Solar radiation studies were initiated in this country in 1911, although the information obtained lacks continuity. At least three projects have been completed recently that indicate radiation levels within the national territory. The first includes an annual radiation map, in addition to monthly radiation maps based on sunshine data compiled in 38 localities. Other meteorological information was then correlated to obtain radiation levels. The second project determined the number of days of insolation by means of photo interpretation of meteorological satellites *Nimbus III* and *ESSA-8* for the period from 1969 to 1971. The last project produced a study of the solar climate of Mexico constituted by insolation information from eight stations in different areas of the country and nine American stations on the border, in addition to information on the duration of insolation from 38 observatories of the Meteorological Service and 98 stations of the Ministry of Water Resources. These studies have made it possible to determine the average daily distribution of solar radiation for each month of the year as well as for the year as a whole.

Solar water heaters have been on the market in Mexico for almost 40 years, and at present some 25 manufacturers produce such equipment. Water desalinization technology has been investigated by a governmental agency and large-scale experimentation has been carried out. Several research centres have built grain-dryer prototypes but none are yet commercially available. Studies on refrigeration using the absorption system are being carried out in two universities. As regards solar electricity, two significant developments have taken place: the production of silicon cells and solar panels, and the production of equipment for the generation of mechanical and electrical energy by means of thermodynamic cycles. Research has been carried out on the passive use of solar energy in housing units.

**Peru**

In Peru, the National Meteorological and Hydrological Service (SENAMI) has electronically computerized information on the number of hours of sunshine and cloud cover per day. This information, which covers the past ten years, has been obtained by means of heliographs located in 84 stations strategically located throughout the country. Nevertheless, no information or publications exist at present on the intensity of solar energy. The Geophysical Institute of Peru and SENAMI have entered into a technical co-operation agreement with the French Government for the preparation of a solar radiation map for Peru. Two pyrometers installed by the Geophysical Institute in June 1977 are operating in the Huayao observatory and on Monte Cosmos. In order to determine overall radiation levels in heliophany stations (number of hours of sunshine), measurements will be taken in 32 stations. Three stations equipped with pyrometers were set up in February 1978 at Ica, Arequipa and Puno to facilitate this research.

The most developed technologies for the use of direct solar energy are those that have been applied to grain dryers, distillation equipment, water heaters, ovens and incubators. Dryers with load capacities of up to 100 kg have been developed locally for the drying of grains and fruit, but are not yet commercially available. Two distillation projects are underway, one for the brackish groundwater in Piura and the other to produce drinking water for goats in Castilla. These technologies are of the labour-intensive type. Water-heating technologies in Peru are quite advanced, both practically and theoretically, and studies have been carried out in this area for more than 50 years. An oven has been developed for the cooking of foodstuffs, in addition to a kiln for firing ceramics. A prototype of a chicken incubator, that functions by storing energy, is also being developed.

### Trinidad and Tobago

The Meteorological Service of Trinidad and Tobago is presently recording the number of hours of daily sunshine. Solar radiation measurements were taken from 1970 to 1973, but later suspended because of deterioration of the instruments. The information collected during this period is available but has not been processed. Furthermore, this information was collected exclusively at Piarcos Airport. Despite the limited information available, solar energy researchers agree that the intensity of solar radiation is more than sufficient for proper functioning of the various types of solar equipment. The highest levels of radiation, recorded during the dry season, range from 550 to 600 Langleys per day, although this figure may drop to between 100 and 150 Langleys per day during the rainy season.

Two groups are presently working on solar energy in the University of the West Indies and prototypes of solar dryers operating by natural or forced convection are being developed. Interesting experiments are also being carried out on the distillation and desalinization of brackish waters and on flat solar collectors. A commercial firm is studying the possibility of local production. Research is being initiated on solar refrigeration and air-conditioning and interest is being shown in local development of solar cells.


## WIND ENERGY

INFORMATION on wind patterns in most of the countries studied has been collected at airports. Although these records do not cover the entire territory of the countries involved, they do provide a general idea of the potential of this resource. Some of the countries studied have been using wind energy for almost 100 years. Its principal use has been and continues to be the pumping of water for domestic use or for supplying water to drinking troughs for pasture animals. The power generated by traditional means (less than 1 kW) is not sufficient to operate pumps for irrigation purposes. Recently, air chargers for batteries have been introduced which are useful for supplying the small quantities of electric power required to operate communications equipment and/or supply domestic lighting requirements.) At present, research projects are being initiated in some countries on the generation of greater amounts of electric power (5-15 kW) for use in small agro-industries.


### Argentina

All existing information has been collected by the National Meteorological Services' observation stations and by the Air Force at airports. These data have not yet been processed and therefore no wind map exists for the entire country.

Traditional, metal multiblade windmills are used extensively in Argentina. Their average power is 0.5 kW and it is estimated that some 500,000 are in operation. These windmills have been manufactured and sold locally for several decades. Medium-powered windmills of from 5 kW to 15 kW are being developed and some already exist as prototypes. The southern part of the country offers prospects for the generation of power and studies are being carried out on the use of large-scale aerogenerators. A wind-based heat generator is being developed for use in Patagonia.

## Brazil

No general map of the country exists indicating average wind velocities. A study was carried out by the Aerospace Technical Centre of San José dos Campos, Sao Paulo, on the northeast coast, where it is estimated that winds are reasonably constant in direction and velocity. Existing information indicates that the best locations for large-scale exploitation of this resource are the southern part of the country and the north-east coast. Electrobras is currently studying the wind patterns on the north-east coast to evaluate their potential.

In almost all regions of this country, water is pumped by windmill energy of less than 1 kW. Windmills are manufactured locally. Scientific and technological research on higher powered aerogenerators (10 kW) is being carried out to develop a prototype. It is estimated that complete medium-powered systems with generators will be put on the market in the near future. Vertical axis generators are being studied in university laboratories.

## Chile and Costa Rica

No wind maps have been prepared in Chile or Costa Rica, although wind velocity is measured at the principal airports. The national meteorological services of these countries have compiled this information.

Certain uses of wind energy have been developed in Chile, particularly for irrigation. Most of the windmills used are manufactured in the country with inexpensive local materials. INTEC has developed a windmill operating at velocities of 6 metres per second and producing a flow of water of 60 cubic metres per day from a depth of 17 metres. Yields of 35 per cent are expected.

In Costa Rica, the University has developed a windmill for the pumping of water. This technology uses imported elements and materials, which has made its production cost extremely high.

## Colombia

HIMAT has recorded and compiled wind velocity data at the country's airports. This organization possesses tables indicating wind velocities and specific times and monthly averages for each of the locations studied. Although these records do not cover the entire country, they do provide a good idea of wind patterns prevailing in each region. The data compiled by HIMAT indicate that one of the most appropriate regions for exploiting wind energy is the Atlantic coast.

The use of wind energy has received a certain amount of attention in this country because of its potentiality in certain regions in which this resource is abundant. An extremely simple and inexpensive cloth-bladed windmill has been developed that produces a yield of water, under normal conditions, of 9 cubic metres per day. A university laboratory has designed, constructed and tested a metal-bladed aerocharger to supply electric power to a rural dwelling. This apparatus produces 500 watts at wind speeds of 7 metres per second. On the industrial level, a local manufacturer is producing an air pump for depths of up to 60 metres and an aerogenerator for charging batteries.

## Guatemala

INSIVUMEH is currently quantifying the velocities, directions and number of hours of wind in 16 stations distributed throughout the country. This information exists for the period from 1973 to 1977. Although this period is short, it has served as a parameter to wind energy researchers for the design of wind-energy equipment. Progress in the use of wind energy has been limited, although certain studies have demonstrated the technical and economic advantage of its use for pumping water. Imported equipment is used in some regions for irrigation and for supplying water to domestic animals.

## Mexico

Information on wind velocities in this country is provided by the National Meteorological Service of the Ministry of Agriculture and Water Resources and is received from 87 meteorological observatories and 3,480 climatological stations throughout the country. From this information, the Energy Sources Division of the Electrical Research Institute has determined the monthly averages over the past two decades of the average velocity of prevailing winds and of maximum velocities. In addition, arithmetical averages per month, per station and per year have been calculated for each of the aforementioned specifics and by climatological station. As of early 1978, information had been processed from 67 meteorological observatories and 96 climatological stations.

Research on windmills is still at the initial stage. Interesting experiments have been carried out with Australian windmills which have demonstrated that this equipment is not appropriate for local conditions. *Instituto de Investigaciones Eletricas* has developed four prototypes of wind equipment: a 6-HP winged aeromotor, a 0.5-HP Savonius-type aeropump, a 200-W Savonius generator and a three-vaned aerogenerator with 1.5 kW aerodynamic section.

## Peru

For thirty years, SENAMI has set up and operated 140 basic meteorological stations that have evaluated wind energy in different regions. These evaluations were made principally with regard to wind direction and velocity as a means of plotting a chart of prevailing winds. In recent years, 60 stations have been established, equipped with anemographs for quantifying wind velocity, direction, current and distribution preparatory to designing equipment to exploit wind energy.

Wind energy has been used for more than two decades for the pumping of water by using windmills of the American type. Dutch windmills of six to eight blades have been produced locally and it is estimated that more than 1,000 are functioning at present. A Savonius rotor system was observed for the pumping of water in desalinization plants in the Piura area. Little progress has been made in the use of wind energy for the generation of electric power or for medium- and high-powered installations.

## Trinidad and Tobago

The Meteorological Service of Trinidad and Tobago has information obtained from wind velocity and direction measurements taken at Piarcas Airport since 1968 and in Matura Airport — in the north-eastern part of the island of Trinidad — since 1974. Both horizontal and vertical axis windmills have been developed by students of the University. Interest in this field has led to training people in the use of this source of energy.

# BIOMASS

FEW COUNTRIES in Latin America have information on the consumption of biomass as fuel and almost none include it in their records or in energy planning. Indirect information on deforestation may serve as indicators but it is obvious that serious assessment efforts must be carried out. Even greater uncertainty prevails with regard to the existence of material to feed anaerobic digestors and generally speaking, no estimates exist on the potential of this resource.

The most important achievement in the use of biomass has been the programme for the production of fuel alcohols derived from sugar cane and cassava carried out in Brazil. The continent's potential for the use of this renewable resource is enormous and may have a far-reaching impact on the environment. As observed in Chapter 2, the generalized use of firewood as a fuel in rural areas has led many countries to serious problems of deforestation and desertification. In such circumstances, conservation measures and programmes for improved use of this resource take on special importance.

## Argentina

No overall evaluation of plant material has been made in Argentina. However, evaluations of forest resources have been made in the northern part of the country where timber-yielding plant material consisting of virgin forest (timber alone) has been found with a yield of 500 cubic metres per hectare and total plant material — including upper branches and dead wood — of 700 cubic metres per hectare.

A project will soon be initiated for the exploitation of sugar cane bagasse. Technology has been developed, although little used, with regard to wood pyrolysis and to advanced and efficient methods for the manufacture of charcoal. A prototype of an anaerobic digestor with a capacity of several cubic metres has been tested for the production of methane.

## Brazil

High levels of solar radiation and humidity, together with a fertile land area, estimated at 70 million hectares, make biomass a potential source of energy and a promising resource. The potential for the production of biomass cannot be easily determined since it is first necessary to determine the kinds of crops that should be cultivated.

Brazil produces almost 40 million metric tons of grain per year and 37.5 per cent of its land area is covered by forests (36 per cent in the Amazon region). Direct use of firewood was 21.3 million tons equivalent of petroleum (MTEP); that of sugar cane bagasse, 4.1 MTEP. Since great efforts are presently being directed at reforestation, it is expected that the production of biogas will increase rapidly over the next decades. Agricultural wastes of 14 harvests in the State of Sao Paulo have been estimated at 14 million metric tons of dry material every year (1976). The production of biogas for energy in Brazil will be of a varied nature owing to different regional characteristics and typical crops. Sugar cane alcohol is produced in large industrial plants in Sao Paulo and Pernambuco; in certain areas of Cerrados, in Minas Gerais and in many other places in the north, eucalyptus and pine are being grown for the production of paper and pulp.

Cassava for the production of alcohol is also being cultivated in Cerrados. The most fertile lands are being used for the production of grains and waste matter from these products has potential use for the generation of energy. Certain typical vegetables of the north-eastern part of the country (babacu, dende) can be used for production of charcoal and oil. (The babacu is a native plant and exists over large areas.)

The alcohol programme has entered its commercial phase. This large-scale programme is providing for utilization of dozens of distilleries, details of which are presented in the case study included in Chapter 7. The use of agricultural waste and wood has been approached in different ways. Steam machines of up to 200 kW power have been developed to burn cereal straw and wood. Substantial technological efforts remain to be exerted in this field and such efforts are being carried out in university laboratories. Research on pyrolysis is still at the laboratory prototype stage and no commercial equipment is available so far. Experiments for the production of methane from urban and agricultural wastes are being carried out in pilot plants in the region of Sao Paulo. Digestors for specific kinds of waste produced in the region are also being studied. Technological adaptation of internal combustion engines for operation with ethanol has been achieved. Prospects for the use of combustible oils and alcohols in Brazil appear promising and may make a significant contribution to solving transportation problems.

## Chile

Chile has the highest percentage in South America of cultivated forest resources in proportion to its land area and it is the country that has made the most detailed study of forest resources. Reliable statistics exist for the entire country.

Chile has good prospects for using plant material for the production of energy. In the 1930s, three million hectares of forest were burned in the region of Aysen. Chile now plans to industrialize this material as charcoal. Great benefits could be derived from the charcoal since Chile imports approximately 3,000 tons of charcoal in briquets every year.

## Colombia

A great deal of firewood is used as a fuel in Colombia. In addition, forests are cut or burned to open up new lands for agriculture. Such depredating use of this resource has brought about serious ecological problems. Statistics show that in 1960 Colombia had 640,000 square kilometres of forest—one-half of the national territory. By 1980, the figure was 350,000 square kilometres, a cause for serious concern. It is estimated that in 1977 total consumption of wood for mines, fences and firewood was of the order of 10 million cubic metres, 55 per cent of which was in the form of firewood. In many areas of the country, firewood is the population's principal energy resource, used mostly for cooking.

Research is being initiated to produce alcohol from tubers which are of scant value as food but rich in starch, such as the bitter cassava and the bire, or from agricultural surpluses. The burning of firewood for cooking food has caused serious deforestation problems in some regions. As a partial palliative to this problem, an oven has been developed that notably increases the efficiency of this source of energy. Official efforts are being made to disseminate the use of this oven since it can be constructed at home. Prototypes exist for the generation of methane from animal and human excrement, mostly derived from foreign experimentation. A prototype anaerobic digestor of coffee pulp is presently being tested.

**Costa Rica**

Costa Rica is suffering from overexploitation of its forest resources, the result of lack of knowledge concerning the resource and the lack of a national policy in this regard. In accordance with the latest agricultural censuses taken in Costa Rica, 60,000 hectares of natural forest are lost every year. If this high rate of deforestation continues, the country's forests will be depleted by the year 2000. A few years ago, the Inter-American Agricultural Sciences Institute estimated that 37 per cent of the total territory of Costa Rica was suitable for forestry, which represents an enormous potential for the use of plant material in this country. An interesting piece of supplementary information is the fact that in 1976, 2.2 million cubic metres of firewood and charcoal were produced. The Instituto Technologico of Costa Rica has constructed a biodigestor for methane and fertilizer production, which is presently in the experimentation stage.

**Guatemala**

In 1950, wood provided most of the energy consumed in the national economy, accounting for 62 per cent of the total; oil accounted for 35 per cent and other sources for the remaining two per cent. At present, these two sources of energy account for almost all the energy consumed, although wood now only represents 33 per cent, whereas oil has risen to 53 per cent. Bagasse accounts for 13 per cent and hydroelectric power for the remaining one per cent. Although the situation has changed considerably since 1950, conservative estimates set the sources of non-commercial energy consumed in Guatemala as 45 per cent of the total. The great importance can thus be seen of non-commercial sources of energy, which are probably the sole source of energy available for more than 60 per cent of the population.

In Guatemala, two significant developments have taken place with regard to the use of plant material:
(1) ***The Lorena Clay Oven:*** This oven is extremely simple to make and can be used for cooking, boiling water and heating the home with any kind of fuel. Initial cost is almost insignificant and it uses only half the fuel consumed by conventional open-fire ovens.

(2) *Pyrolysis,* or thermal decomposition of material containing carbon in the absence of oxygen, has been studied locally and experimentation has begun in the installation of pyrolysis plants. Experimentation in the production of biogas began in Guatemala in 1953. Since then, uninterrupted progress has been made in digestor technology. At present, an industrial plant is being constructed with a capacity of 8,000 cubic metres per day of methane gas.

## Mexico

According to the Mexican Oil Institute's publication *Energeticos: panorama actual y perspectivas,* the consumption of vegetable fuels in Mexico has decreased notably over the past decades. According to data from the Economic Commission for Latin America, in 1940 it represented 15 per cent of the total gross consumption of energy in Mexico; in 1950, this share was reduced to 8 per cent and by 1955, only 6 per cent. In 1972, according to estimates made by the Department of Economic Studies and Industrial Planning of the Mexican Oil Institute, the consumption of vegetable fuels —used mainly in the home — accounted for approximately 3.9 per cent of national energy consumption. As shown in the information contained in the same publication, this reduction should be understood in relative terms since consumption of vegetable fuels has diminished very little. In 1960, vegetable fuels were consumed by 22.7 million inhabitants — 65 per cent of the population. In 1970, they were consumed by 21.1 million inhabitants, 44 per cent of the population at that time. Total consumption was estimated at 4.5 million tons of charcoal and firewood (at the rate of 212 kg per person per year) which is equivalent to 2.1 million cubic metres of crude oil (13.2 million barrels). If this trend continues, in 1982, some 19.5 million inhabitants (26 per cent of the population forecast for that year) will consume vegetable fuels with an energy equivalent of 1.8 million cubic metres of crude oil (11.3 million barrels).

Although the consumption of vegetable fuels continues at high levels in Mexico and represents serious ecological consequences, no significant research has been identified on optimum use of plant materials. Several institutions are studying the production of methane gas by means of anaerobic digestion of organic waste. A communal digestor was recently put into operation with a capacity of 35 cubic metres per day.

## Peru

It was possible to ascertain that 40 per cent of the population lives in the mountainous areas of the Andes, which accounts for less than 20 per cent of the country's entire area. Because of the topography, distribution of energy is extremely difficult; therefore, large amounts of firewood are used for cooking and heating homes in an area that is cold all year round because of the altitude. Further information concerning plant material consumption in Peru may be obtained from *Balance Nacional de Energia*. Similarly, no significant research has been identified to improve the use of plant material for energy. Research is beginning in some institutions on the generation of biogas.

**Trinidad and Tobago**

It was not possible to obtain information as to whether accurate studies exist on forest resources or the magnitude of deforestation in this country. Apparently this problem has not been studied so far, although it is known that in 1976, Trinidad and Tobago produced 10,000 cubic metres of firewood and charcoal. Research is being carried out on anaerobic fermentation of sugar cane bagasse and construction of a prototype digestor is planned.

## HYDROPOWER

A DRAINAGE AREA may be considered as the product of two groups of factors: climatological and physiographical. The first includes rain and evaporation; the second, the characteristics of the basin and the river bed. The drainage area is composed of two parts: a basic flow, rising from underground drainage, and a direct drainage, produced by the rain.

There is a series of statistical and empirical methods for calculating drainage, although direct measurement is always preferred; the latter is done in most countries but limited only to flows that offer possibilities of large power generation (usually higher than 10 MW). Very little is known of flows for small-scale generation. A few parameters that may offer a general view of the water resources in the countries under consideration can be given: the yearly average precipitation and its space distribution, the average basin yield — that is, the average volume or streaming flow generated in a specific soil area, and the seasonal rain distribution, which is of great importance specifically in small basins since these are the ones where microgeneration may not occur.

The possiblity of generating electricity through the use of small waterfalls (installed capacities of between 25 kW and 1 MW) has aroused increasing interest in most of the countries visited. Technologies exist of limited sophistication that would make it possible to supply small communties with electricity and solve problems related to communications, domestic comfort and agro-industries for the processing and conservation of agricultural products.

A new approach to rural electrification is being applied in this field that differs from the traditional approach of supplying electric power from enormous power stations by means of extensive distribution networks. Technological progress has been made and an official project in Colombia for the installation of 20 small-scale power plants is discussed in Chapter 7.

## Argentina

Two regions may be distinguished according to registered fluvial precipitation: the first to the north of the Colorado River and the second to the south of the same water flow. In the first region, the average annual precipitation decreases from north-east to south-east from 1,500 mm to 400 mm; in the second, this average annual precipitation decreases from west to east, from 2,500 to 200 mm. Two-thirds of the country has an annual average precipitation lower than 500 mm; these are arid or semi-arid areas. The average water yield in the country is 7.8 $l/s/km^2$. In regard to the monthly variation of precipitation, the Quiaca region presents the greatest of differences, since in December and January precipitations register 67 mm and 87 mm, respectively. From April to August, they are practically nil. On the other hand, there are precipitations almost throughout the year.

National manufacturers have produced equipment for the generation of electric power by means of small waterfalls for the past 40 years. At present, turbines of the conventional type are being manufactured—Kaplan (without turbine wickets), Francis and Pelton. Intake conduits, velocity regulators and electrical generators are also produced locally. One hundred and thirty small-scale power plants exist in Argentina with average capacities of 500 kW. Mitchell-type turbines are being tested and electronic velocity regulators are being manufactured.

## Brazil

The average annual precipitation in the northern region (Amazon Basin) exceeds 2,500 mm. In most parts of the southern region, rainfall reaches approximately 1,000 mm a year. In the coastal region of the north-east, rain exceeds 1,000 mm a year, while in the central region it varies between 500 mm to 1,111 mm. In some regions, rain is less than 500 mm a year, with monthly irregularity. The average water yield of basins is 26 $l/s/km^2$. The monthly variations are rather pronounced in several regions of the country (for example, in the region of Brasilia, 292 mm of rain are registered in December and 0 mm in June, July and August; in Vaupes, the lower precipitations are 127 mm in September).

Small-scale generation systems are being manufactured that combine diverse flows with diverse waterfalls. They have not been put into generalized use so far, but there are great prospects for this equipment in the near future. At present, feasibility studies are being carried out for the State of Amazonas and the Territory of Rondonia.

## Chile

In the north of Chile, 105 mm average rainfall has been registered. Rain rapidly increases towards the south, where 141 mm in La Serena, 500 mm in Valparaiso and 2,707 mm in Valdivia have been registered. Average yields are extremely variable; thus, while in Copiapo the yield shows 0.4 $1/s/km^2$, in Tolten it is 102 $1/s/km^2$. In general, monthly variations are minimum. No projects have been identified for the use of small waterfalls during the course of this study.

## Colombia

In order to describe the average annual precipitation, four regions can be defined. Firstly, the Caribbean coastal region, where an average of 1,143 mm of rain per year is registered. Secondly, the central plateau region, where rain reaches 3,500 mm in some areas. Thirdly, the Pacific lowlands, where rain reaches beyond 5,000 mm per year; and fourthly, the eastern region, with 635 mm a year. In regard to the monthly variation of rain, it can be stated that in Andagoya the least difference occurs since the lowest value recorded is 487 mm in December and the maximum is 645 mm in June. In Cartagena, precipitations are nil in February.

The small-scale power station project of the Colombian Electric Power Institute (ICEL) is the most important being carried out for the exploitation of small waterfalls. These studies are an attempt to decide on 30 locations for the installation of power stations with a capacity ranging from 100 kW to one MW of electric power. *Colciencias* has financed studies on the use of Colombian water resources.

Colombia is a country with great hydroelectric potential, not only for the construction of large-scale projects but for the generation of small capacities as well. The latter have been the subject of substantial research activities. Small plants using Mitchell-type turbines have been set up with capacities of 0.5 to 3 kWe for domestic use in isolated areas. Studies are being made to develop power stations in the 5 to 30 kWe range, for which turbines and dams have been designed. Chapter 7 provides a detailed account of a governmental project for the construction of some 20 small-scale plants with capacities of 100 KWe to 1 MWe. Local developments include civil works projects and turbines. No electrical generators are manufactured locally.

## Costa Rica

There are abundant precipitations in most of the territory. The annual average is 7,000 mm. In the driest areas, such as Valle Central and Provinces of Guancaste and Pantanera, the average annual precipitation is 1,400 mm. The average water yield is 59 $1/s/km^2$. Two well-defined seasons appear and the registered variations in the fluvial precipitations range from 5 mm in February to 300 mm in

September. As numerous rivers in Costa Rica have already been studied in detail, the possibility of using small waterfalls for the generation of energy appears to be promising. No projects were identified for the use of small waterfalls during the course of this study.

## Guatemala

The rainy season starts in May and usually ends in October. In the region of El Peten, there is an average annual precipitation of 2,000 mm. In Cuchumates, in the region of Sierra Madre, 3,000 mm of rainfall are registered. In some zones, rainfall reaches less than 500 mm. The average yield in the basins is 28 $l/s/km^2$. Monthly variations amount from 2.5 mm in February to 270 mm in June in the nearby region of Guatemala City.

## Mexico

A great part of the northern region is arid or semi-arid and rainfall is registered to be lower than 500 mm a year. Rainfall in Mexico City is 574 mm average and in Puebla, annual rainfall is 890 mm. The states having higher precipitations are Veracruz and Tabasco, where seasonal variations are also presented. The average annual yield is 6 $l/s/km^2$. Monthly variations are high in most of the country: for example, in La Paz, precipitations hardly exist during the year except in August and September, when precipitations are registered at 30 mm and 35 mm, respectively.

The Institute of Electrical Research (IIE) has made a study of small-scale hydroelectric power systems of 5 to 100 kW capacity. This study has shown that there are approximately 150 small hydroelectric plants installed in Mexico. Eighty per cent of these plants are in the States of Chiapas, north of the city of Tapachula, in the municipalities of Coahuatan, Motozintla, Huixtla and Tapachula, whereas the remaining 20 per cent are located principally in the States of Veracruz in the municipality of Coatepec. A few installations are also located in the states of Oaxaca, Jalisco and Tamaulipas. These small-scale power plants are used principally in the coffee plantations in the states of Chiapas and Veracruz. The IIE's study shows that of the 21,000 isolated communities in Mexico of a population of 150 to 500 inhabitants, some 2,000 have small streams or rivers with sufficient year-round flow and head to obtain water power for the production of electric or mechanical energy.

One hundred and fifty small-scale power plants are operating in this country. Some of them were installed at the beginning of the century, the most recent during the 1950s, principally on coffee plantations. Small workshops are engaged in producing turbines. In the Instituto de Investigaciones Electricas, research has been carried out in this field, with tests coupled to an overshot wheel and a Mitchell Banki turbine.

## Peru

Space distribution of rainfall is variable; thus, in some areas practically no rain falls, as is the case in Pisco, while in the area of Quincemil, rainfall surpasses 5,000 mm a year. Areas of the Selva have the following distribution: north, 3,000 mm; centre, 4,000 mm; and south, 5,000 mm average rainfall a year. The average annual rainfall in the country is 2,500 mm, the average annual yield being 5 $1/s/km^2$. Monthly variations are high; for instance, in Arequipa, from April to October rainfall is nil, and in February, it reaches 45 mm. In Cuzco, the variation is registered to be between 160 mm in January and 5 mm in June and July.

Peru has a great potential for the installation of small-scale power plants. Projects are being initiated for the technological development of turbines, basically in ITINTEC and in ELECTROPERU; the latter has begun a construction programme of 40 microgenerators.

## Trinidad and Tobago

Average annual rainfall is 1,605 mm. Its monthly variation is 40 mm in February and 242 mm in August. No projects were identified for the use of small waterfalls during the course of this study.

A significant outcome of the field visits made during the course of this study was evaluation of the degree of progress noted with regard to equipment or technology observed and to the institutions working on such technologies. Tables 3.1 to 3.10 provide a list of the national institutions working on different subject areas and the degrees of technological progress they have achieved. Table 3.10 provides an overall summary indicating:

    (a) — the degree of advance in technology with regard to various non-conventional sources of energy in each country;

    (b) — the trends or preferences in technological development in those countries;

    (c) — where technologies exist commercially for the use of non-conventional sources of energy;

    (d) — which technologies of non-conventional sources of energy have not been developed or are still in a barely incipient stage.

From this section onwards, a series of percentage values will be given in parentheses; they are the results of a survey at regional level which (without being representative since the size of sampling is considered unsatisfactory) provide the order of magnitude and, in some cases, an approach for a quantitative support to the criteria given.

## Table 3.1  Development of Technologies for exploiting Non-conventional Sources of Energy in Argentina

| ARGENTINA | 1 | 2 | 3 | 4 | 5 | 6 | 7 | 8 | 9 | 10 | 11 | 12 | 13 | 14 | 15 | 16 |
|---|---|---|---|---|---|---|---|---|---|---|---|---|---|---|---|---|
| **Direct Solar Energy** | | | | | | | | | | | | | | | | |
| Hot air drying | | ◨ | | | | ◨ | | | | | | | | | | |
| Distillation | | | | | | | | | | | | | | | | |
| Water heating (incl. intermediate temp.) | ◸ | ◨ | ◸ | | | | | ◨ | | | | ◨ | | | | |
| Solar kitchens and ovens | ◸ | | | | | | | | | | | | | | | |
| Solar refrigeration | | | | | | | | | | | | | | | | |
| Solar electricity (photothermal) | | | | | ◸ | | | | | | | | | | | |
| Evaluation of resource | | | | | | | | | | | ◸ | ◸ | | | | |
| **Wind Energy** | | | | | | | | | | | | | | | | |
| Windmills for pumping water | | | | | | | | | | | | | ■ | | | |
| Aerochargers | | | ◸ | | | | | | | | ◸ | | ■ | ◸ | | |
| Evaluation of resource | | | | | | | | | | | | | | ◸ | | |
| **Use of Plant Material** | | | | | | | | | | | | | | | | |
| Firewood and charcoal | | | | | | | | | | | | | | | ◨ | |
| Briquets | | | | | | | | | | | | | | | | |
| Alcohols and combustible oils | | | | | | | | | | | | | | | | |
| **Generation of Biogas** | | | | | | | | | | | | | | | | |
| Digestors | | | | | | | | | | | ◸ | | | | ◨ | |
| **Exploration of Small Water Flows and Small Waterfalls** | | | | | | | | | | | | | | | | |
| Small-scale power plants | | | | | | | | | | | | | ■ | | | |
| Pumps and hydraulic rams | | | | | | | | | | | | | | | | |
| Water mills | | | | | | | | | | | | | | | | |
| Dam construction | | | | | | | | | | | | | | | | |
| **Machinery and Infrastructure for Utilizing Alternative Sources of Energy** | | | | | | | | | | | | | | | | |
| Halloarchitecture | ◸ | | ◨ | ◨ | | | | | | | | | | | | |
| Pedal-driven machines | | | | | | | | | | | | | | | | |
| Others | | | | | | | | | | | | | | | | |

Does not exist □   Research carried out ◸   Prototype exists ◨   Commercialized ■

Institutions

(1) National University of Salta
(2) National University of Rosario
(3) National Atomic Energy Commission (CNEA)
(4) Argentine Arid Zones Institute
(5) National University of San Luis
(6) National Commission for Geohallophysical Studies
(7) National University of Luján
(8) National University of Tucumán
(9) Technical Institute of Buenos Aires
(10) National Meteorological Service
(11) National Space Research Commission
(12) Patagonian Centre
(13) Various Industries
(14) TOMECO (Mendoza), SMAR (Buenos Aires)
(15) National Agricultural Technology Institute
(16) National Industrial Technology Institute

## Table 3.2 Development of Technologies for exploiting Non-conventional Sources of Energy in Brazil

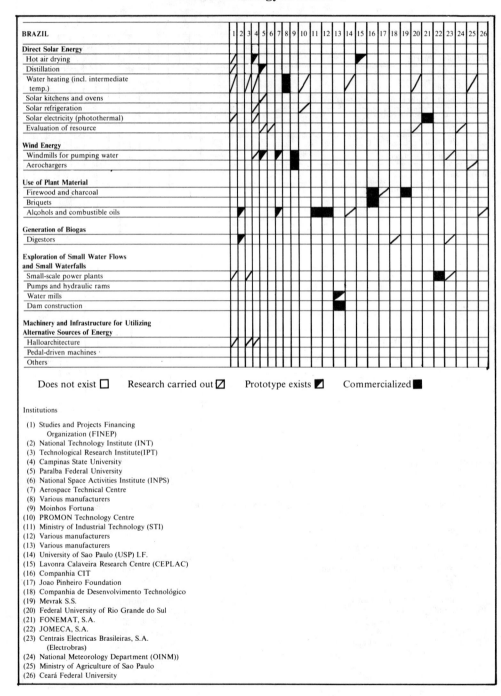

Does not exist □   Research carried out ◨   Prototype exists ◪   Commercialized ■

Institutions

(1) Studies and Projects Financing Organization (FINEP)
(2) National Technology Institute (INT)
(3) Technological Research Institute (IPT)
(4) Campinas State University
(5) Paraiba Federal University
(6) National Space Activities Institute (INPS)
(7) Aerospace Technical Centre
(8) Various manufacturers
(9) Moinhos Fortuna
(10) PROMON Technology Centre
(11) Ministry of Industrial Technology (STI)
(12) Various manufacturers
(13) Various manufacturers
(14) University of Sao Paulo (USP) I.F.
(15) Lavonra Calaveira Research Centre (CEPLAC)
(16) Companhia CIT
(17) Joao Pinheiro Foundation
(18) Companhia de Desenvolvimento Technológico
(19) Mevrak S.S.
(20) Federal University of Rio Grande do Sul
(21) FONEMAT, S.A.
(22) JOMECA, S.A.
(23) Centrais Electricas Brasileiras, S.A. (Electrobras)
(24) National Meteorology Department (OINM))
(25) Ministry of Agriculture of Sao Paulo
(26) Ceará Federal University

# RENEWABLE SOURCES OF ENERGY IN LATIN AMERICA

## Table 3.3    Development of Technologies for exploiting Non-conventional Sources of Energy in Chile

| CHILE | 1 | 2 | 3 | 4 | 5 | 6 |
|---|---|---|---|---|---|---|
| **Direct Solar Energy** | | | | | | |
| Hot air drying | | P | | P | | P |
| Distillation | | | | | | |
| Water heating (incl. intermediate temp.) | | C | C | | | |
| Solar kitchens and ovens | | | | | R | |
| Solar refrigeration | | | | | R | |
| Solar electricity (photothermal) | | | | | R | R |
| Evaluation of resource | | | | | R | |
| **Wind Energy** | | | | | | |
| Windmills for pumping water | | P | | | | |
| Wind transportation | | | | | | |
| **Use of Plant Material** | | | | | | |
| Firewood and charcoal | | P | | | | |
| Briquets | | P | | | | |
| Alcohols and combustible oils | | | | | | |
| Algae | | | P | | | |
| **Generation of Biogas** | | | | | | |
| Digestors | | P | | | | |
| **Exploration of Small Water Flows and Small Waterfalls** | | | | | | |
| Small-scale power plants | | | | | R | |
| Pumps and hydraulic rams | | | | | | |
| Water mills | | | | | | |
| Dam construction | | | | | | |
| **Machinery and Infrastructure for Utilizing Alternative Sources of Energy** | | | | | | |
| Halloarchitecture | | | R | | | R |
| Pedal-driven machines | | | | | | |
| Others | | | | | R | |

Does not exist ☐    Research carried out ◨    Prototype exists ◩    Commercialized ■

Institutions

(1) INTEC
(2) LIRQUEN
(3) University of Chile
(4) Federico de Sta. Maria Technical University
(5) CORFO
(6) State Technical University—Chile

ENERGY ALTERNATIVES IN LATIN AMERICA

**Table 3.4   Development of Technologies for exploiting Non-conventional Sources of Energy in Colombia**

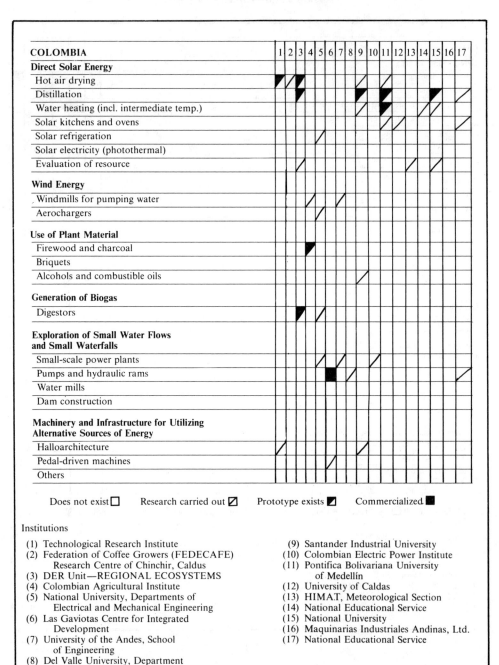

Does not exist ☐   Research carried out ◪   Prototype exists ◤   Commercialized ■

Institutions

(1) Technological Research Institute
(2) Federation of Coffee Growers (FEDECAFE) Research Centre of Chinchir, Caldus
(3) DER Unit—REGIONAL ECOSYSTEMS
(4) Colombian Agricultural Institute
(5) National University, Departments of Electrical and Mechanical Engineering
(6) Las Gaviotas Centre for Integrated Development
(7) University of the Andes, School of Engineering
(8) Del Valle University, Department of Physics
(9) Santander Industrial University
(10) Colombian Electric Power Institute
(11) Pontifica Bolivariana University of Medellín
(12) University of Caldas
(13) HIMAT, Meteorological Section
(14) National Educational Service
(15) National University
(16) Maquinarias Industriales Andinas, Ltd.
(17) National Educational Service

## Table 3.5 Development of Technologies for exploiting Non-conventional Sources of Energy in Costa Rica

| COSTA RICA | 1 |
|---|---|
| **Direct Solar Energy** | |
| Hot air drying | |
| Distillation | |
| Water heating (incl. intermediate temp.) | ◪ |
| Solar kitchens and ovens | ◪ |
| Solar refrigeration | |
| Solar electricity (photothermal) | |
| **Wind Energy** | |
| Windmills for pumping water | ◪ |
| Aerochargers | |
| **Use of Plant Material** | |
| Firewood and charcoal | |
| Briquets | |
| Alcohols and combustible oils | |
| **Generation of Biogas** | |
| Digestors | ◪ |
| **Exploration of Small Water Flows and Small Waterfalls** | |
| Small-scale power plants | |
| Pumps and hydraulic rams | |
| Water mills | |
| Dam construction | |
| **Machinery and Infrastructure for Utilizing Alternative Sources of Energy** | |
| Halloarchitecture | |
| Pedal-driven machines | |
| Others | |

Does not exist □     Research carried out ◪

Prototype exists ◪     Commercialized ■

Institution

(1) University of Costa Rica,
    School of Engineering,
    School of Mechanical and Electrical Engineering

## Table 3.6 Development of Technologies for exploiting Non-conventional Sources of Energy in Guatemala

| GUATEMALA | 1 | 2 | 3 | 4 |
|---|---|---|---|---|
| **Direct Solar Energy** | | | | |
| Hot air drying | | ◪ | | |
| Distillation | | | | |
| Water heating (incl. intermediate temp.) | ◪ | | | |
| Solar kitchens and ovens | ◪ | | | |
| Solar refrigeration | | | | |
| Solar electricity (photothermal) | | | | |
| **Wind Energy** | | | | |
| Windmills | ◪ | | ◣ | |
| Wind transportation | | | | |
| **Use of Plant Material** | | | | |
| Firewood and charcoal | ■ | | | |
| Briquets | | | | |
| Alcohols and combustible oils | | ◪ | | |
| Pyrolisis | | ◪ | | |
| **Generation of Biogas** | | | | |
| Digestors | ◣ | | | |
| **Exploration of Small Water Flows and Small Waterfalls** | | | | |
| Small-scale power plants | | | | |
| Pumps and hydraulic rams | | | | |
| Water mills | | | | |
| Dam construction | | | | |
| **Machinery and Infrastructure for Utilizing Alternative Sources of Energy** | | | | |
| Halloarchitecture | | | | |
| Pedal-driven machines | | | | |
| Others | | | | |

Does not exist ☐  Research carried out ◪

Prototype exists ◪  Commercialized ■

Institutions
(1) Mesoamerican Appropriate Technology Centre
(2) Central American Industrial Research and Technology Institute
(3) Phillip School
(4) Local technology

**Table 3.7** Development of Technologies for exploiting Non-conventional Sources of Energy in Mexico

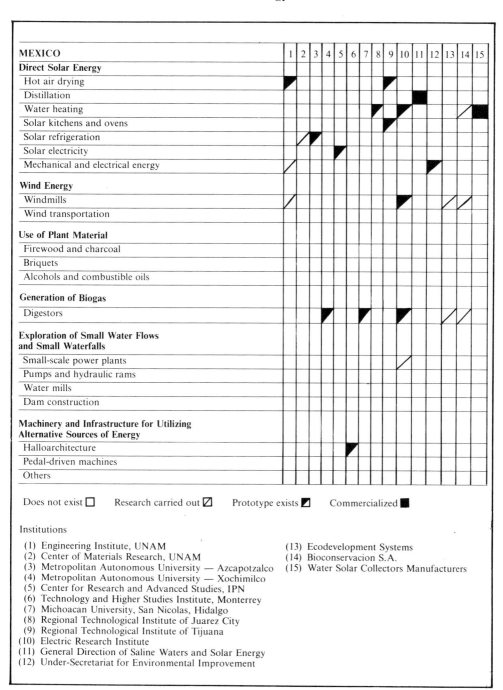

Table 3.8  Development of Technologies for exploiting Non-conventional Sources of Energy in Peru

| PERU | 1 | 2 | 3 | 4 | 5 | 6 |
|---|---|---|---|---|---|---|
| **Direct Solar Energy** | | | | | | |
| Hot air drying | | ◪ | | | | |
| Distillation | | ◪ | | | | |
| Water heating (incl. intermediate temp.) | | ◪ | ◪ | | ■ | |
| Solar kitchens and ovens | | ◪ | ◪ | | | |
| Solar refrigeration | | | | | | |
| Solar electricity (photothermal) | | | | | | |
| Evaluation of resource | | | ◪ | | | ◪ |
| **Wind Energy** | | | | | | |
| Windmills | | ■ | | ■ | | |
| Aerochargers | | | | | | |
| **Use of Plant Material** | | | | | | |
| Firewood and charcoal | | | | | | |
| Briquets | | | | | | |
| Alcohols and combustible oils | | | | | | |
| **Generation of Biogas** | | | | | | |
| Digestors | | | | | | |
| **Exploration of Small Water Flows and Small Waterfalls** | | | | | | |
| Small-scale power plants | | ◪ | | | | |
| Pumps and hydraulic rams | | | | | | |
| Water mills | | | | | | |
| Dam Construction | | | | | | |
| **Machinery and Infrastructure for Utilizing Alternative Sources of Energy** | | | | | | |
| Halloarchitecture | | | | | | |
| Pedal-driven machines | | | | | | |
| Others | | | | | | |

Does not exist □         Research carried out ◪

Prototype exists ◪       Commercialized ■

Institutions

(1) Institute for Industrial Technological Research and Technical Standards (ITINTEC)
(2) National Engineering University, Institute of Applied Natural Energy, San Cristobal de Humanga National University
(3) La Molina Agrarian University
(4) Local technology (Arequipa)
(5) Ernesto Bareda
(6) Andahuaylas Higher Technical Institute

## Table 3.9  Development of Technologies for exploiting Non-conventional Sources of Energy in Trinidad and Tobago

| Trinidad and Tobago | 1 | 2 | 3 | 4 | 5 |
|---|---|---|---|---|---|
| **Direct Solar Energy** | | | | | |
| Hot air drying | ■ | ■ | | | |
| Distillation | ■ | ■ | | | |
| Water heating (incl. intermediate temp.) | ■ | ■ | | | |
| Solar kitchens and ovens | | ◪ | | | |
| Solar refrigeration | | ◪ | | | |
| Solar electricity (photothermal) | | | | ◪ | |
| **Wind Energy** | | | | | |
| Windmills for pumping water | | | ◪ | | |
| Wind transportation | | | | | |
| **Use of Plant Material** | | | | | |
| Firewood and charcoal | | | | | |
| Briquets | | | | | |
| Alcohols and combustible oils | | | | | |
| **Generation of Biogas** | | | | | |
| Digestors | | | ◪ | | ◪ |
| **Exploration of Small Water Flows and Small Waterfalls** | | | | | |
| Small-scale power plants | | | | | |
| Pumps and hydraulic rams | | | | | |
| Water mills | | | | | |
| Dam construction | | | | | |
| **Machinery and Infrastructure for Utilizing Alternative Sources of Energy** | | | | | |
| Halloarchitecture | | | | | |
| Pedal-driven machines | | | | | |
| Others | | | | | |

Does not exist ☐    Research carried out ◪    Prototype exists ◪    Commercialized ■

Institutions
(1) University of West Indies, Chemistry Department
(2) University of West Indies, School of Engineering
(3) University of West Indies, Chemical Engineering Department
(4) Caribbean Industrial Research Institute
(5) Government of Trinidad

**Table 3.10   Non-conventional sources of energy and their applications**

| Country | | | Argentina | Brazil | Chile | Colombia | Costa Rica | Guatemala | Mexico | Peru | Trinidad and Tobago |
|---|---|---|---|---|---|---|---|---|---|---|---|
| Solar | a | Hot Air Drying | R | C | C | C | | | C | C | |
| | b | Distillation | R | C | C | | | | C | C | |
| | c | Water Heating | C | C | C | C | | | C | C | |
| | d | Solar Kitchens and Ovens | R | C | C | | | | C | C | |
| | e | Solar Refrigeration | R | P | P | | | | C | R | |
| | f | Solar Electricity | R | C | | C | | | C | C | |
| | g | Inventory of Resource | I | I | I | I | I | | I | I | I |
| Wind | a | Windmills for Pumping Water | C | C | C | C | | | C | C | R |
| | b | Aerogenerators | R | C | R | C | | | C | R | |
| | c | Inventory of Resource | I | I | I | I | I | | I | I | I |
| Plant Material | a | Firewood and Charcoal | C | C | C | C | | C | C | | |
| | b | Briquets | R | C | R | | | | | | |
| | c | Alcohols and Combustible Oils | C | C | R | | | | R | | |
| | d | Inventory of Resource | I | I | I | I | I | | I | I | I |
| Biogas | a | Digestors | C | C | C | C | | | C | R | I |
| Small Waterfalls | a | Small-scale Power Plants | R | C | R | | | | | R | |
| | b | Pumps and Hydraulic Rams | | C | R | | | | | R | |
| | c | Water Mills | | C | R | | | | | | |
| | d | Dam Construction | | C | C | | | | | | |
| | e | Inventory of Resource | I | I | I | | | | | I | |

Does not exist □   Research carried out ▨   Prototype exists ▧   Commercialized ■   Inventory of Resources exists ⊠

CHAPTER 4

# Industrial Capacity and Human Resources in Latin America for Producing Equipment to Harness Renewable Sources of Energy

## INDUSTRIAL CAPACITY

IT WOULD BE of great interest to evaluate the Latin American region's industrial capacity for producing the equipment required to exploit renewable sources of energy. That, however, would imply detailed analysis that the time available for the present study did not permit. Consequently, it was necessary to make an evaluation by means of somewhat imprecise methods. On the basis of overall industrialization data rather than on particular studies by country, the installed capacity was estimated for various branches of industry that might eventually produce equipment for the exploitation of renewable sources of energy. The industries studied were heavy machinery, electricity, electronics, chemicals and construction.

The following procedure was used to obtain information on the region's potential for producing such equipment:

(1) Equipment of greatest potential use was studied and broken down into its component parts or subsystems.
(2) After obtaining information on equipment components, the industrial processes were analyzed that are required both for manufacture of components (raw materials and manufacturing as such) and for the production of equipment from the components.
(3) Once information about the manufacturing involved was obtained, it was possible to determine whether or not these industrial processes exist at present in the installed industrial capacity of each of the countries studied.

For the purposes of this evaluation, the countries studied were divided into three categories according to their degrees of industrial development (Table 4.1).

Table 4.2 shows the components into which the equipment using renewable sources of energy was divided; Table 4.3 shows the industrial processes involved in their manufacture. Table 4.4 classifies equipment according to its degree of sophistication. Analysis of the information collected led to the general conclusion that all countries in the region have sufficient installed industrial capacity to

### Table 4.1 Classification of Countries according to their relative degrees of Industrial Development

| Countries | Relative Degree of Industrial Development | Code |
|---|---|---|
| Argentina, Brazil, Mexico | High | A |
| Chile, Colombia, Peru | Medium | B |
| Costa Rica, Guatemala, Trinidad and Tobago | Low | C |

manufacture technologically less sophisticated equipment.*

It should be noted, however, that the level of sophistication of the equipment or processes involved is also a relative notion, since the industrial processes known in some countries of the region are non-existent in others. Generally speaking, technologies associated with the production of electricity and the manufacture of liquid fuels are of a high degree of sophistication, so that it is precisely in these fields that the countries of lesser industrial development will encounter the greatest difficulties. Technologies associated with the use of wind energy and the generation of biogas are usually of intermediate sophistication and, although installed industrial capacity exists in all countries, design and materials technology problems may still be encountered.

On the other hand, technologies associated with the use of direct solar energy for heating, distillation and drying are at low levels of sophistication. This does not mean that little remains to be accomplished with regard to design and the study of materials; nevertheless, all the countries visited could successfully undertake large-scale production of such equipment.

Solar refrigeration is a relatively sophisticated technology in which technical and design problems continue to occur even on a worldwide level. It is not unreasonable to suppose that in the near future the region will be at a stage of industrialization for producing such equipment, since a great deal of research and development remains to be carried out in this area. In summary, Table 4.3 provides an evaluation of the region's industrial capacity for producing equipment to use renewable sources of energy.

It does not appear that a lack of industrial capacity is to blame for Latin America's failure to develop its own capacity to manufacture this equipment. Rather, it appears that the reasons for this lack are principally due to problems of technology or design. Therefore, emphasis in future should be placed on design and the solution of technological problems rather than on the installation of new productive processes.

* We wish to express special appreciation to the Nacional Financiera of Mexico and UNIDO's project *Capital Goods* for the collaboration and information provided.

**Table 4.2  Components system**

| Type of Energy | Components |
|---|---|
| *Direct solar energy* | |
| Hot air drying | Collector; drier reservoir |
| Distillation | Tray with polluted water, condensing glass, reservoir |
| Water heating | Collector, water circuit, hot and cold water tanks |
| Solar kitchens and ovens | Collector, concentration screens, energy storehouse, food reservoir |
| Solar refrigeration | Collector, heat exchanger, refrigeration chamber |
| Solar electricity | Photocells or gas expanders, turbine, generator, pump |
| *Wind energy* | |
| Windmills for pumping water | Rotor, transmission, pump |
| Aerogenerators | Blades, generator, alternator, batteries |
| *Use of plant material* | |
| Firewood and charcoal | |
| Briquets | |
| Alcohols and combustible oils | Distillation systems |
| *Generation of Biogas* | |
| Digestors | Intake, digestion chamber, gas outlet, effluent outlet, tributary, measurement and analysis equipment |
| *Exploitation of small water flows and small waterfalls* | |
| Small-scale power plants | Civil engineering works, piping, generator, turbines |
| Pumps and hydraulic rams | |
| Water mills | Wheel, transmission |
| Dam construction | Materials, locks, valves |

Table 4.3  Manufacturing processes

*Direct Solar Energy*
- Hot air drying — Sheet metal work, basic chemistry, mechanical engineering/glass
- Distillation — Sheet metal work, mechanical engineering, basic chemistry
- Water heating — Sheet metal work, mechanical engineering, electrical chemistry
- Solar kitchens and ovens — Sheet metal work, mechanical engineering, chemistry
- Solar refrigeration — Sheet metal work, mechanical engineering, chemistry, boilermaking
- Solar electricity — High purity chemistry, precision mechanics

*Wind Energy*
- Windmills for pumping water — Mechanical engineering, casting, machining, ball bearing, laminating, stamping
- Aerogenerators — Stamping, fibreglass, electrical

*Use of Plant Material*
- Firewood and charcoal — Mechanical engineering
- Briquets — Mechanical engineering
- Alcohols and combustible oils — Chemistry, boilermaking, electrical

*Generation of Biogas*
- Digestors — Boilermaking, electrical, mechanical engineering

*Exploitation of Small Water Flows and Small Waterfalls*
- Small-scale power plants — Casting, boilermaking, mechanical engineering, electrical
- Pumps and hydraulic rams — Casting, boilermaking, mechanical engineering
- Water mills — Casting, boilermaking, mechanical engineering
- Dam construction — Casting, boilermaking, mechanical engineering, construction

# INDUSTRIAL CAPACITY AND HUMAN RESOURCES

**Table 4.4  Sources of Energy and their Applications**

| COUNTRY | Solar a — Hot Air Drying | Solar b — Distillation | Solar c — Water Heating | Solar d — Solar Kitchens and Ovens | Solar e — Solar Refrigeration | Solar f — Solar Electricity | Wind a — Windmills for Pumping Water | Wind b — Aerogenerators | Plant Material a — Firewood and Charcoal | Plant Material b — Briquets | Plant Material c — Alcohols and Combustible Oils | Biogas a — Digestors | Small Waterfalls a — Small-scale Power Plants | Small Waterfalls b — Pumps and Hydraulic Rams | Small Waterfalls c — Water Mills | Small Waterfalls d — Dam Construction |
|---|---|---|---|---|---|---|---|---|---|---|---|---|---|---|---|---|
| A — Argentina | ⊠ | ⊠ | ⊠ | ⊠ | ⊠ | ■ | ⊠ | ⊠ | ⊠ | ⊠ | ⊠ | ⊠ | ⊠ | ⊠ | ⊠ | ⊠ |
| A — Brazil | ⊠ | ⊠ | ⊠ | ⊠ | ⊠ | ■ | ⊠ | ⊠ | ⊠ | ⊠ | ⊠ | ⊠ | ⊠ | ⊠ | ⊠ | ⊠ |
| B — Chile | ⊠ | ⊠ | ⊠ | ⊠ | ⊠ | ■ | ⊠ | ⊠ | ⊠ | ⊠ | ■ | ⊠ | ⊠ | ⊠ | ⊠ | ⊠ |
| B — Colombia | ⊠ | ⊠ | ⊠ | ⊠ | ⊠ | ■ | ⊠ | ⊠ | ⊠ | ⊠ | ■ | ⊠ | ⊠ | ⊠ | ⊠ | ⊠ |
| C — Costa Rica | ⊠ | ⊠ | ⊠ | ⊠ | ■ | ■ | ⊠ | ■ | ⊠ | ⊠ | ■ | ⊠ | ⊠ | ⊠ | ⊠ | ⊠ |
| C — Guatemala | ⊠ | ⊠ | ⊠ | ⊠ | ⊠ | ■ | ⊠ | ⊠ | ⊠ | ⊠ | ⊠ | ⊠ | ⊠ | ⊠ | ⊠ | ⊠ |
| A — Mexico | ⊠ | ⊠ | ⊠ | ⊠ | ■ | ■ | ⊠ | ■ | ⊠ | ⊠ | ⊠ | ⊠ | ⊠ | ⊠ | ⊠ | ⊠ |
| B — Peru | ⊠ | ⊠ | ⊠ | ⊠ | ⊠ | ■ | ⊠ | ⊠ | ⊠ | ⊠ | ⊠ | ⊠ | ⊠ | ⊠ | ⊠ | ⊠ |
| C — Trinidad and Tobago | ⊠ | ⊠ | ⊠ | ⊠ | ⊠ | ■ | ⊠ | ⊠ | ⊠ | ⊠ | ■ | ⊠ | ⊠ | ⊠ | ⊠ | ⊠ |

**Key:**

Inventory of Resources exists

A — Countries with relatively high degrees of development
B — Countries with medium degrees of development
C — Countries with relatively low degrees of development

■ No installed industrial capacity exists
⊠ Installed industrial capacity exists

## HUMAN RESOURCES

A STUDY of capacities should include considerations on the human resources required for the development of such capacity. Consequently, it would be appropriate to postulate reasonable prospects for the use of renewable sources of energy and determine *a posteriori* what human resources the region will require to carry out such projects effectively. Unfortunately, this has not been done in the past since this study reveals that whatever exists in this field at present has been developed with the assistance of the professionals available in each country, the great majority of whom have not been specially trained to work in this subject area.

The specialities most concerned with the use of renewable energy include various branches of engineering and basic sciences: mechanical engineering (45 per cent), physics (13 per cent), electrical engineering (9 per cent), chemistry (5 per cent), geology (5 per cent), industrial engineering (3 per cent), civil engineering (2 per cent). There is also the participation of specialists in applied sciences (related to the environment, architecture, rural areas and agricultural production) as well as in social sciences and economics. This study has shown that all these fields are involved to a greater or lesser degree in analyzing situations in which the use of renewable sources of energy shows promising prospects and would contribute to solving important problems of the population. The academic level of the researchers is acceptable in most cases, with a high percentage of project directors (16 per cent) having done post-graduate studies in different universities.

One of the problems that projects on renewable energy sources face is the short period of time that the researchers dedicate to the studies. In general, the number of full-time researchers is low (29 per cent); the remainder (40 per cent) are people who spend one to 25 per cent of their time on these projects. What is surprising, however, is the relatively small number of scholarship fellows participating in these projects (12 per cent). This does not meet one of the objectives of research and development, namely, forming a staff of able personnel in the different areas of Renewable Energy Sources.

Although the study reveals enormous differences in the quality and quantity of people available in the countries studied, these countries (with the exception of Mexico) have made no systematic efforts to provide training in this field in universities or advanced study institutions. The only efforts identified in this direction were isolated research or seminars. Although it is difficult to generalize, the impression remains that the curriculum of professionals being trained is traditional and that scientific research activities and technological development itself ultimately place students in contact with the particular problems of energy and its relationship to development.

It has been found that this subject has been dealt with in many of the universities and institutes visited, however not always with the necessary depth. Furthermore, the scant tradition of scientific research prevailing in many institutions in the region leads them to formulate over-ambitious and somewhat unrealistic plans. For example, a different programme is often assigned to each institution, which is obviously incompatible with the in-depth study required. Lastly, it should be noted that the lack of economic support often seriously limits such studies. Tools are required to be able to measure, compare, judge and improve.

The following are the major problems affecting Renewable Energy Sources: a lack of budget (45 per cent); the lack of instrumentation, equipment and laboratories (26 per cent); a shortage of personnel, both scientific and technical (23 per cent); and the lack of specialized bibliography (20 per cent). Apparently, these problems could be solved locally: they could be solved within each country. However, the problem must be examined in the regional context of Latin America. In almost every case, the technology under consideration already exists in other countries (80 per cent) and is generally adapted (70 per cent), even though only a few institutions (20 per cent) maintain experience, technology and personnel exchanges with their foreign counterparts. Moreover, in the majority of cases (63 per cent), no marketing studies exist of the technologies that individual countries of the region intend to adopt.

These facts lead to the conclusion that the usefulness of Renewable Energy Sources in Latin America decreases considerably, due to the constant repetition of research (a typical example being the development of the domestic solar collector), as well as the lack of an intraregional exchange and marketing studies to support current investigations, specifically in the cases of adaptation. A fictitious situation exists, therefore, with regard to the technological development since it generally does not reach the application stage, merely remaining in the laboratory constituting the pilot centres.

To increase the usefulness of Renewable Energy Sources and to assist in diminishing the problems cited (even though they are not the only ones), a technological information network that would be operational at the regional level must be established. The advantages of these technologies are national production of equipment (77 per cent); the use of local labour (77 per cent); reduction of the costs of equipment (55 per cent); and the possession of national patents (38 per cent).

Nevertheless, there has been a recent proliferation of projects based on appropriate technology, also known as soft technologies. Some of these projects are concerned with technologies for the use of renewable sources of energy. However, they do not reveal any systematic effort to make a detailed study of the subject nor are they generally associated with training professionals. Instead, these projects concentrate on technologies having little hope of fulfilling needs not covered by conventional sources of energy, particularly in rural areas. The region does have personnel of sufficient quality and quantity to deal with and solve problems involving less sophisticated technologies for the use of renewable sources of energy; however, such personnel cannot hope to undertake research in the near future involving high levels of sophistication nor, in some cases, even medium levels of sophistication.

The generation of solar electricity, the production and use of hydrogen, the design of special motors that can consume non-traditional fuels, the study of powerful electric aerogenerators and the production of alcohols and combustible oils are examples of technologies to be developed in the future that require personnel with high-level, specialized scientific and technical training. It is also true that many countries, particularly the smallest or least developed countries, cannot tackle this energy challenge individually nor would it make sense for them to attempt to do so. Thus, Europe's example should be followed with regard to large-scale programmes (such as the aerospace and nuclear programmes) in which joint efforts are made through the conclusion of agreements to amass greater capacity. In the same manner, conditions for possible and fruitful co-operation should be created in Latin America, both with regard to technology and the training of personnel. It appears necessary to set up advanced institutes, specialized in particular areas in which professionals from all countries could receive sound and thorough training.

One cannot disregard the fact that advanced countries are working actively to solve the energy problem with sophisticated and modern approaches. This might create a new dependence on the part of Latin America in the future similar to its dependence on energy technology up to the present. The only appropriate way to avoid this problem—and allow creative and efficient technological pluralism to prevail in the region—is to initiate immediately the systematic training of scientists and professionals who are required to carry out adequate research on the region's true energy requirements.

CHAPTER 5

# Marketing Renewable Sources of Energy Equipment

### TECHNOLOGY AND PENETRATION

DISCOVERY of a technology does not ensure that it will be put to large-scale or commercial use. Many examples may be cited of the long and difficult process of transforming invention into a commercial product. This process also depends on an enormous variety of economic, social and political factors;* consequently long periods of time—in some cases dozens of years—may be required to carry such processes to successful conclusions. For example, the internal combustion engine was invented on an experimental basis many years before the manufacture of the first automobile and many more years before automobiles began to be produced commercially. Conversion of the automobile into a universally used product fully incorporated into our lifestyle also required many years.

In the field of renewable sources of energy, the problem at hand is to transform technological innovations into commercial equipment and have them become absorbed by diverse social groups to the greatest extent possible. This does not mean that the development of renewable sources of energy should be sought exclusively. On the contrary, development of these sources should be aimed at solving specific problems in specific places in order to achieve genuine technological pluralism in the energy sector.

### FACTORS DETERMINING THE PENETRATION PROCESS

THE PENETRATION process of technologies using renewable sources of energy is determined by diverse factors that variously influence these different kinds of technology. The following discussion presents a list of these factors, which, for presentation purposes, are classified as technological, economic, social, cultural and political. In practice, all are closely interrelated.

---

\* The distinction between the functions of invention and innovation was first postulated by Schumpeter: innovation enables an invention to be transformed into a product or service useful to society.

## Technological Factors

**Design:** An important aspect of the design of equipment is its simplicity of use and maintenance, since such equipment is normally used by people of relatively low educational level and in isolated places where technical services do not exist for supervising the functioning of the equipment or making any required repairs. Since users value convenience, equipment must satisfy the need for energy with the least effort and concern on the part of the user. This is precisely one of the reasons consumers prefer electric power, although the cost per kilowatt hour is higher than that of other energy sources. The convenience element, for example, has led to solar collectors often being manufactured with automatic equipment to operate electrically when the temperature drops below a certain point. Manufacturers who have not included this feature have lost a certain amount of competitiveness.*

Related to simplicity of design is the autonomy of the equipment, its capability of being used independently of other forms of energy. This is an important factor for certain uses. In some cases, design should consider specific characteristics of the places in which the equipment is to be used. Such is the case of solar collectors in areas with very hard water where it may be necessary to use different materials.

**Quality:** The quality, maintenance and reliability of equipment greatly influence its degree of acceptance by potential users. Cases have been known of unscrupulous manufacturers who have sold equipment of poor quality, insufficient duration and deficient operation to the detriment of the penetration of such equipment. Strict quality control is therefore extremely important. In some countries, considerable importance is being given to this aspect: for example in Argentina, the Argentine Solar Energy Association (ASADES) plans to set up a testing bank; in Brazil, quality guarantee certificates accompany such equipment. In most cases the question of useful life and economic viability of equipment is a problem of materials technology that obliges countries either to import materials or to expend considerable research efforts to improve domestic materials.

**Availability of the resource:** Certain factors limit either the degree of penetration inherent in the resource itself or in its technological development. Obviously, one of these factors is the intermittency of the resource, which makes it necessary to find economical methods for its accumulation or to limit its use to those who do not require constant availability of energy.

**Level of Technological Development:** In other cases, limitation derives from the fact that technology has not progressed sufficiently to make the manufacture of

---

* This is the reason why a certain portion of the public in Mexico continues to buy solar collectors imported from the United States despite the fact that they are also manufactured domestically.

equipment commercially viable. This is the case, for example, with regard to solar energy to produce high temperatures of 250° C or more.

It may also happen that manufacturers have not made efforts to make changes in their equipment that would definitely increase their penetration. For example, wind-chargers could have a relatively larger market if it were possible to increase their power, which does not appear to be too difficult from a technical standpoint.*

Unfortunately, with regard to research in Latin America, scientists are usually reluctant to perform research on soft technology, preferring to devote their efforts to more complex and sophisticated problems associated with concerns of the scientific community in the industrialized countries. This is compounded by a lack of contact between universities and the production sector, which means that research is rarely directed toward solving concrete problems— in this case the energy problems—of large sectors of the population.**

## Economic Factors

It has been stated previously that the economic viability of equipment is a necessary, although not sufficient, condition for penetration of a technology. We will now analyze the factors that contribute towards increasing the economic viability of equipment.

**Production Cost:** The determining factor here appears to be the production cost of equipment, although some researchers interviewed felt that this was a secondary factor, since substantial support is required from government to be able to introduce such equipment on the market. However, this argument is not valid since subsidies or any other kind of incentives are ultimately paid for by the entire community; consequently, if mass consumption is desired, the equipment must be produced at the lowest possible price.

**Economy of Scale:** It is important to bear in mind that a kind of vicious circle exists with regard to cost since one of the important variables determining the cost of a specific piece of equipment is the volume of production, which in turn is determined by the degree of penetration of the technology, further influenced by the production cost of equipment that uses the technology. Overcoming this vicious circle requires clear-cut policies to support the development of the energy sector.

**Operation Cost:** The economic viability of equipment and consequently the penetration of a technology are strongly influenced by the operation cost of alternative equipment, which is principally determined by the cost of conventional

---

* Interview with Mr Pablo Brosens, partner of the SRL agro-industrial firm in Buenos Aires.

** Interview with a salesman from the Leopard, SA firm, manufacturers of windmills.

energy. The cost of energy for users is a political variable that depends essentially on the rate policies of the government concerned. If gas or electricity are inexpensive—as is the case in Trinidad—it is highly unlikely that equipment using renewable sources of energy will be competitive—except, of course, in areas of the country that are not well supplied with conventional energy.* Such dependence of the price of energy on political factors makes it possible for the situation to change suddenly, so that equipment that is competitive today may cease to be so tomorrow and *vice versa*.

**Costs of Transfer of Technology:** The cost of equipment using renewable sources of energy manufactured under license requiring the payment of royalties or patents may be higher in accordance with the arrangements made for the transfer of the technology in question. In countries with balance of payments problems, the situation may be aggravated by the scarcity of foreign exchange.

**Intensity of Use of Equipment:** Another factor that often makes it difficult for equipment to be economically viable is inadequate use. Such is the case with cereal driers which are only used for a few months in the year. An example of this may be found in the Salta region of Argentina, where an attempt has been made to manufacture solar tobacco driers. However, in order to encourage farmers to use them, some alternative use must be found that will make it possible to use them for longer periods; otherwise an inordinate amount of time is required to amortize their cost.

**Marketing Network:** Another factor having an influence on the penetration of such equipment is the need for an adequate marketing network that will make it possible to inform rural inhabitants of the possibilities offered by the use of non-conventional sources of energy and to have them observe how such equipment functions without the need to travel to urban centres. An adequate marketing network would also make it possible to provide appropriate technical servicing to the users of such equipment.

**Distribution of Income:** The distribution of income in a given country is another factor that influences the prospects for the penetration of such equipment since, in addition to other factors, the size of the market is also important. On several occasions during the course of the interviews, it was noted that solar collectors at present can only be acquired by middle- and high-income sectors.

**Existence of Support Infrastructure:** The use of certain non-conventional sources of energy requires support infrastructure to facilitate the work of the user. For example, the development of a programme of alcohol fuel would not be viable if it did not include provisions for supply stations that would fulfil the function of the

---

* A manufacturer of solar collectors in Argentina said his firm had decided not to attempt to introduce this equipment in the city of Buenos Aires since it was difficult to compete with gas heaters because the city was well supplied with relatively cheap and efficiently distributed gas.

gasoline stations presently supplying spent fuels. Establishment of such an infrastructure would probably require heavy investment and, therefore, support from the state is essential.

## Social Factors

**Resistance to Innovations in the Rural Sector:** Technological innovations in the industrial sector usually encounter little resistance since, within the framework of economies of scale, their essential objective is to increase profits, which is relatively easy to achieve. In the rural sector, however, the situation is different since it is much more difficult to perceive the relationship between the innovation involved and the increase in production or improvement in the quality of life that may be obtained. In addition, a general lack of confidence exists with regard to innovations.

**Resistance to Innovations in the Public Sector:** Resistance to new technologies also exists in the public sector. Certain government officials regard the use of renewable sources of energy with great scepticism, since such innovations are not to be found within the framework of technological pluralism and are consequently considered almost second-class technologies to which countries seeking development cannot assign great importance.

**Structural Factors:** Land tenure structures and the degree of organization of small farmers are other factors that influence the penetration of these technologies. Landless *campesinos* or small landowners have no energy alternative other than human or animal energy or that of depredating the soil. Obviously in such cases, the possiblity of using other energy sources arouses great interest. Certain equipment using renewable sources of energy are commercially justified if they can produce an amount of energy larger than the requirements of a single family. For this reason, a certain degree of *campesino* organization is essential in order to be able to use some of this equipment jointly.*

## Examples of Cultural Factors

**Cooking and Eating Habits of Users:** This factor should also be considered in designing solar energy equipment, particularly solar kitchens. An example of how the failure to consider this aspect has been detrimental to the penetration of solar kitchens—one that is often criticized in the literature on this subject—is the case of a design for kitchens to be installed in the open air that was not accepted by the

---

* Experience has been accumulated in India regarding the use of individual or collective biogas plants. Many of the difficulties encountered are not of a technical nature but rather are derived from lack of community organization.

community because housewives were accustomed to cooking inside their homes. The case also exists of kitchens using biogas, which were also rejected because the food cooked in them bore a taste to which the users were unaccustomed.

**External Status Symbols:** In some communities the possession of certain equipment or implements associated with urban domestic life may constitute status symbols, for example gas or electric stoves instead of wood-burning stoves, or conventional water heaters. This situation may constitute an additional difficulty for the penetration of certain equipment in some communities.

**Political Factors**

**Government Attitudes:** The penetration of this equipment is substantially determined by governmental policies. To date, the development of renewable sources of energy has generally been looked upon with little interest by governments since these sources will produce only long-term results and governments are under heavy pressure to deal with short-term problems.

A determining factor in government attitudes is their development concepts. Governments that include among their principal development objectives that of attempting to satisfy the basic needs of the population more likely will give much greater priority to renewable sources of energy than those whose main objective is to increase the per capita income growth rate.

**Attitudes of Certain Groups:** At times, economic interests among certain groups of society may encourage them to frustrate the development of renewable sources of energy. During the Seminar on Energy in Guaruja, Brazil, the instance was mentioned of certain agricultural co-operatives that had opposed the introduction of solar soybean driers since it was felt that they would provide greater negotiating power to *campesinos* who would no longer feel pressured to sell this product.*

**Attitudes of Large Corporations:** During the same seminar, several participants were of the opinion that in some cases opposition from large transnational corporations operating in the field also existed, since the multiplication of decentralized systems would diminish the great power they yield at present.

---

* Address by Professor Meyer of the University of Campinas.

# EXTENT OF PENETRATION OF DIVERSE TECHNOLOGIES USING RENEWABLE SOURCES OF ENERGY

TABLE 5.1 provides a general idea of the extent of penetration of certain technologies in the various countries included in this study. This chart shows present penetration that could be achieved under optimum conditions when the principal factors limiting such penetration have been removed. Present and potential penetration are classified for each country and for each technology according to four relative levels. Level 0 indicates zero penetration, with no possibility of such technology being developed in the country. Level 1 indicates a relatively slight degree of penetration; level 2, a medium degree of penetration; level 3, a high degree.

Classification of the different technologies according to various degrees of penetration is essentially a subjective evaluation based on observations made during visits to the countries and obtained from the opinions of the persons interviewed in each instance. Consequently, they are to be considered more as impressions rather than a classification based on quantitative criteria.

To obtain more objective figures would require detailed analysis of each of the degrees of penetration of the various technologies with regard to the various social groups. Overall penetration would then be determined by the weighted average of all the sectoral penetrations.

Potential penetration should be defined for one or several determined periods of time: although some of the factors determining penetration are independent of time, others, such as technological development, will be modified over the years. Table 5.1 shows projections for potential penetration over the next 10 or 15 years based on an assumption of the degree of development that the various technologies will have attained in that time. The difference between the present level of penetration and the potential level gives an idea of the technologies that in the space of 10 or 15 years may be used much more intensively if proper policies are put into practice.

## Solar Energy

Solar energy, together with wind energy, is clearly the renewable source of energy that offers the greatest possibilities for the future and that has currently achieved the greatest degree of development. The equipment of greatest potential is constituted by solar heaters and driers. These combine, on the one hand, greater technological development and, on the other, the fact that they would satisfy an evident need in the rural areas of Latin America.

Distillation equipment can achieve a medium degree of penetration in Chile, where interesting experiments in desalinization have been carried out for years; in Trinidad and Tobago, a research group has built and installed several units of this type, which may prove to be useful for demonstration purposes. Other solar

Table 5.1  Extent of Penetration of Renewable Sources of Energy in different countries

| Source | Technology | Argentina | | Brazil | | Chile | | Colombia | | Costa Rica | | Guatemala | | Mexico | | Peru | |
|---|---|---|---|---|---|---|---|---|---|---|---|---|---|---|---|---|---|
| | | A | P | A | P | A | P | A | P | A | P | A | P | A | P | A | P |
| Direct Solar | Hot air drying | 0 | 2 | 0 | 3 | 0 | 1 | 0 | 2 | 0 | 2 | 0 | 2 | 0 | 3 | 0 | 1 |
| | Distillation | 0 | 1 | 0 | 1 | 1 | 2 | 0 | 1 | 0 | 1 | 0 | 1 | 0 | 1 | 0 | 1 |
| | Water heating | 1 | 2 | 1 | 2 | 1 | 2 | 0 | 2 | 0 | 1 | 0 | 1 | 1 | 3 | 0 | 2 |
| | Solar kitchens and ovens | 0 | 0 | 0 | 1 | 0 | 0 | 0 | 0 | 0 | 1 | 0 | 0 | 0 | 1 | 0 | 1 |
| | Refrigeration | 0 | 0 | 0 | 1 | 0 | 0 | 0 | 0 | 0 | 0 | 0 | 0 | 0 | 1 | 0 | 0 |
| | Solar electricity | 0 | 1 | 0 | 1 | 0 | 0 | 0 | 0 | 0 | 1 | 0 | 0 | 0 | 1 | 0 | 0 |
| Wind | Windmills | 2 | 3 | 2 | 3 | 0 | 2 | 0 | 2 | 0 | 2 | 0 | 2 | 0 | 2 | 0 | 2 |
| | Aerogenerators | 2 | 3 | 1 | 2 | 0 | 2 | 0 | 2 | 0 | 2 | 0 | 2 | 0 | 3 | 0 | 2 |
| Plant Material | Firewood/Charcoal | 0 | 1 | 1 | 2 | 0 | 1 | 0 | 1 | 0 | 0 | 1 | 2 | 1 | 2 | 1 | 2 |
| | Briquets | – | – | 1 | 2 | – | – | – | – | – | – | – | – | – | – | – | – |
| | Alcohols/Oils | 0 | 1 | 1 | 2 | 0 | 0 | 0 | 0 | 0 | 1 | 0 | 1 | 0 | 1 | 0 | 0 |
| Biogas | Digestors | 0 | 1 | 0 | 2 | 0 | 0 | 0 | 0 | 0 | 1 | 1 | 3 | 0 | 1 | 0 | 0 |
| | Small-scale power plants | 0 | 1 | 0 | 2 | 0 | 2 | 1 | 2 | 1 | 2 | 1 | 2 | 1 | 2 | 1 | 2 |
| Small Waterfalls | Pumps | – | – | – | – | – | – | – | – | – | – | – | – | – | – | – | – |
| | Watermills | – | – | – | – | – | – | – | – | – | – | – | – | – | – | – | – |
| | Dams | – | – | – | – | – | – | – | – | – | – | – | – | – | – | – | – |

A = Present extent of penetration;   P = Potential extent of penetration (within 10 to 15 years);   0 = None;   1 = Slight;   2 = Medium;   3 = High equipment is of much greater technological complexity, so that present and potential penetration is slight or totally non-existent.

## Wind Energy

An interesting tradition in the use of wind energy exists in several countries of the region, especially in Argentina and Brazil. Incipient technological development efforts will doubtless lead to a significant increase in penetration and, consequently, potential penetration in most of the countries has been estimated as high.

## Plant Material

The use of this energy resource is quite intensive in the region, although present penetration levels have been low in some cases, since such use is carried out in an anarchical and depredatory manner.

In several countries, such as Brazil and Guatemala, research projects exist that are designed to make greater use of plant material. As the technology used for these purposes is not very complex, it may be assumed that higher penetration will take place in the future.

The Central American energy programme has defined as its main priority a project for the manufacture of alcohol fuel. Consequently, these countries, including Brazil (which has also concentrated efforts in this area), show a higher penetration potential than other countries.

## Biogas

The only current experience of any importance in this field is that of Guatemala. For this reason, it has a penetration rating of 1, whereas in all the other countries it is 0. The experience acquired by Guatemala will also signify a high degree of penetration in the future.

## Small Waterfalls

Generally speaking, this source of energy has a high potential degree of penetration; however, its present use and penetration are relatively modest.

According to the survey carried out by several researchers of the region, solar energy is the renewable source with the greatest penetration possibilities towards 1985 (27 per cent of the opinion); next is wind energy (23 per cent), followed by small waterfalls harnessing (15 per cent), methanol production (8 per cent), biogas (4 per cent) and geothermia (4 per cent).

# ECONOMIC SECTORS THAT COULD OPT FOR RENEWABLE SOURCES OF ENERGY TECHNOLOGIES

As shown in Table 5.2, the rural sector is the one that can best opt for the use of Renewable Sources of Energy since, with the exception of methanol and geothermia, the option percentages of its use of other sources are high. Table 5.2 also shows that solar energy is the one with the greatest application in each of the economic sectors.

## Policies directed towards Increasing the Level of Penetration

Almost all of the studies that have been made on the possibility of increasing the use of renewable sources of energy agree that an essential requisite in so doing is Government support that can be translated into implementing the policies formulated for this purpose. This opinion was repeated on many occasions by people interviewed during the present study.

The support the State must provide for the development of these energy sources cannot be indiscriminate. It must be centred on sources that use the natural resources possessed by the country in question and are most adaptable to its social and economic realities. Feasibility studies must be carried out in order to select those technologies with the most potential for producing a positive impact on the economy.*

Projects related to the development of renewable sources of energy cannot be evaluated simply on a private cost basis since the positive effects of such projects benefit the community as a whole. Consequently, project evaluation methods should be used that take into consideration the social, environmental and other effects such projects may bring about—methods such as cost-benefit analysis and the methodology recently developed by Technology Assessment. Cost-benefit analysis offers the advantage of not considering private costs or benefits but rather social ones, which are reflected in their opportunity cost.

It is interesting to note the conclusions of an evaluation of biogas plants using this methodology (Barnett, 1976). These conclusions may be applied to other renewable sources of energy. The principal conclusion is that biogas plants are most feasible in the following situations: (1) when diverse inputs have a low opportunity cost, that is when they have practically no alternate use; (2) when the

---

* In this regard, awareness is being created on both national and international levels. For example, the Agency for International Development (AID) plans to initiate a support programme for extensive feasibility studies to determine future lines of renewable sources of energy.

**Table 5.2** Option percentage: economic sectors that may opt for the utilization of NCES

| Sector<br>Source | Urban Domestic | Industrial | Transport | Services | Transformation of Energy | Rural |
|---|---|---|---|---|---|---|
| Solar | 25 | 25 | | 10 | 10 | 30 |
| Wind | 16 | 16 | | 8 | 16 | 56 |
| Biogas | | | | | 25 | 75 |
| Biomass Combustion | | 33 | | | 33 | 33 |
| Geothermy | | 50 | | | 50 | |
| Small Waterfalls | | 14 | | | 28 | 58 |

**Source:** Survey on New Energy Sources in Latin America

efficiency of plant operation is adequate and (3) when the alternatives to the products manufactured have a high opportunity cost.

These conclusions are also applicable to other renewable sources of energy, and consequently the State should support their development in situations when this is most feasible.

With respect to the first situation, an analysis is needed of what alternative uses there are of capital, labour, natural resources and other inputs that are employed to produce energy from renewable sources. Normally, natural resources (sun, wind, organic wastes) have no alternative uses; however, in some cases they do. One instance that demands careful study is the production of alcohol fuel from agricultural products, principally from sugar cane. In this case, the problem arises whether one can justify using land resources to produce energy instead of food.

In the second situation, it should be noted that the study of plant efficiency is essentially technical. In some cases, efficiency may be increased by using hybrid systems, such as fossil fuels plus solar energy. For example, using solar collectors as fuel 'economizers' may be much more attractive than using complete systems based on this kind of energy.

Lastly, the third situation depends on the opportunity cost of conventional energy. Here, it is important not to consider the price of the fuel which may be subsidized, but rather its opportunity cost—the real cost in resources of having that fuel available for the country in question.

Once it has been determined which technologies using renewable sources of energy would have the most positive effects by being developed, the State should assume responsibility for carrying out whatever policies are necessary to foment such development—in other words, to increase penetration. Policies oriented towards increasing penetration may be of diverse types and, for the purposes of study, may be classified according to the objectives they aim to fulfil.

## Policies directed towards Promoting Technological Development of Equipment Using Renewable Sources of Energy

The State may make a decisive contribution to the development of technologies concerning renewable sources of energy by assisting in financing research and encouraging such research to solve problems of a practical nature. This may be accomplished within the general framework of governmental scientific and technological policy. Governments must attempt to establish some kind of control over the quality of equipment to ensure that it complies satisfactorily with the function for which it is designed. Current levels of technological knowledge are relatively high and it consequently appears more appropriate to devote greater efforts to seeking practical applications of already-known technology than to develop new technologies. Efforts should be concentrated on improving certain technological parameters—such as windmill power or obtaining higher temperatures by means of solar energy—which would make it possible to increase the market for such equipment.

## Policies directed towards Increasing the Economic Viability of Equipment using Renewable Sources of Energy

In this respect, several parallel tactics may be adopted as a means of achieving, *inter alia,* the following objectives:

(a) diminishing the cost of production of equipment: this may be achieved by means of furnishing technical and economic support to producers either by assisting them in acquiring inputs or by co-operating in their efforts to increase productivity;

(b) expanding the market: a few measures that may contribute to achieving this objective are:
—purchase of equipment by the public sector;
—financial support to users to assist them in making the initial investment;
—elimination of legal restrictions that in some countries prevent the free use of renewable sources of energy.

(c) increasing the intensity of use of the equipment: this will make it possible to reduce the amortization period;

(d) co-operating in the organization of a marketing network;

(e) increasing the competitiveness of equipment using renewable sources of energy: this may be achieved by means of exemptions or incentives designed to reduce the operating cost of new equipment, although this may be difficult since this cost normally is in itself quite low. Another possibility is to attempt to maintain the price of renewable energy as close as possible to its opportunity cost, which generally means increasing its price. A more drastic alternative consists of directly restricting the use of conventional energy by limiting the sale of fuel or in some way establishing strict environmental controls.

## Policies directed towards Diminishing Resistance to Innovation

An important role may be played by campaigns oriented towards obtaining greater acceptance of innovations by both users in rural areas and government officials on various levels.

Generally speaking, there is a lack of information available concerning the real advantages and limitations of the use of renewable sources of energy as compared with the use of conventional sources. This lack of information, which must be remedied, limits penetration.

Existing agricultural extension services may be useful during information campaigns. Other instruments may also be used, such as the construction of pilot plants or equipment as a means to demonstrate the advantage offered by the use of renewable sources of energy.

## Policies directed towards Considering Cultural Factors

It is most important for the technicians—who are responsible for designing equipment to use renewable sources of energy—to understand cultural limitations clearly and to adapt the characteristics of such equipment to cultural realities. Often the problem is presented improperly by attempting to see how the cultural patterns of eventual users may be modified so that they will accept the technology in question. In reality, the matter should be dealt with by attempting to determine how cultural factors may be considered in designing such equipment. These policies may only be implemented insofar as governments fully assume the role they can play in the development of renewable sources of energy.

## References

Barnett, Andrew (1976) *Social and Economic Evaluation of Biogas Plants.* University of Sussex.

CHAPTER 6

# Institutional Aspects of the Use of Renewable Sources of Energy in Latin America

## CONDITIONING FACTORS IN THE USE OF RENEWABLE SOURCES OF ENERGY IN LATIN AMERICA

ON THE BASIS of the considerations presented so far, it is felt that governments would do well to determine accurately the present capacities of their countries to use renewable sources of energy, while at the same time establishing institutional machinery and defining policies to increase this capacity rapidly. The problem of supplying energy to the population in Latin America has been approached by establishing institutional organizations of a diverse nature. Thus, countries have strengthened or established public energy corporations. In 1950, public production of electricity amounted to approximately 10 per cent of the total, whereas in 1974 it amounted to 78 per cent. Moreover, there is a trend towards stepping up state activities in all stages of the oil and natural gas industries and governments are steadily extending their control over oil-refinery corporations.

Capacity to use renewable sources of energy indicates that countries are in a position to:

(a) determine the amount of energy resources existing in their various geographic areas;

(b) have available a 'critical mass' of scientists and researchers with knowledge of the technical advances achieved in this field in various parts of energy to the fullest extent possible and take advantage of the diversity of specific applications;

(c) design and produce equipment at such cost and quality as to be able to transform these renewable sources into usable energy — thermal, electrical or mechanical — to satisfy the basic needs of the population;

(d) promote the most widespread distribution and commercialization of such equipment;

(e) provide extensive publicity to all sectors of the population regarding the possibilities and advantages offered by renewable sources of energy (this factor is of particular importance in rural areas where there is less possibility of obtaining energy from other sources);

(f) provide financial support to ensure implementation of the above requirements.

Effective fulfilment of these conditions can only be possible to the extent that appropriate institutional machinery is available and the policies necessary to develop these sources of energy are put into practice. Furthermore, institutional machinery would have to provide support to researchers, producers, distributors and users of equipment utilizing renewable sources of energy.

The following division into subject areas is somewhat arbitrary, since these areas are closely interrelated and consequently have reciprocal influence on one another.

## ENERGY, ENERGY POLICY AND RENEWABLE SOURCES OF ENERGY

THE DEVELOPMENT of renewable sources of energy demands singular awareness and concern on the part of those who make decisions in the energy sector in order to translate such decisions into well-defined policies. It is important that explicit policies —those laid down officially for the sector — should not contradict implicit policies — those defined by other sectors that impinge upon the energy sector. It may happen, for example, that the explicit policy calls for reducing the consumption of fossil fuels; however, such a policy will have little chance of success if, at the same time, implicit policies exist that contradict this objective, such as a policy entailing artificially low fuel prices or a transportation policy based fundamentally on the development of highway transportation. For such concordance between energy and the policies promoted by other sectors to exist, some kind of high-level machinery must also exist with genuine authority to enforce established policies and resolve any differences that may arise with other sectors.

From an institutional standpoint, it is important to stress that the energy sector in Latin America has been characterized by direct mediation by governments in the form of autonomous, state-affiliated corporations and by ministries and the like, either alternatively or complementarily. Normally, the debate among those responsible for this sector has revolved about thermal plants and hydroplants, although nuclear energy has more recently entered the picture. Practically no importance has been given to renewable sources of energy, which some even consider to be second-class technologies incapable of providing energy in the quantities required. Only recently have these sources been included in energy planning in some countries, although with relatively secondary concern. In practice, this has generally meant that a special working group has been appointed to study the possibilities of these sources.

In recent years, national energy commissions have been established in almost all the countries of the region, with the objective of making comprehensive studies of the energy problem. These commissions have varying degrees of authority and in practice have functioned with disparate results. Nevertheless, they are an important first step, since the possibility of developing a renewable source of energy should be one of the most important topics to be considered in commissions of this kind.

Symptomatically, those responsible for the energy sector are beginning to assign more importance to these sources of energy by providing a place for them in institutional structures. For example, in Argentina a working group was recently set up on a renewable source of energy that is now in the information-collecting stage. It is expected that during the second stage of the energy plan (from 1985 to 2000) the explicit role to be reserved for renewable sources of energy will be included. The National Energy Committee was established in Costa Rica in July 1976 as 'an organization to assist the Ministry of Economy, Industry and Commerce in defining policies to be followed with regard to the energy problem'.* In that country, a National Commission for Saving Energy also exists that, in a report (1976) on the energy situation, proposed, *inter alia,* to 'encourage the use in certain areas of other sources of energy, such as wind energy, heat energy provided by firewood and charcoal, etc.' and 'to promote the use of solar energy as a partial substitute for electric power for the heating of water for domestic and industrial use.'

It is also important to take into account the regional and subregional organizations that have been set up with the primary objective of dealing with the energy situation. For obvious reasons, they have concentrated on non-renewable sources of energy; however, they may also serve as valuable instruments to encourage the development of renewable sources of energy.

Among such organizations is the Latin American Energy Organization (OLADE), which in August 1978 held a seminar on energy economy in which renewable sources were considered. An interesting experience in the region is the Central·American energy programme, which initiated its second phase in 1978 and whose general objective is 'to contribute to the socio-economic development of the Central American countries by setting down the basis for a comprehensive policy for planned development of energy' (UNDP,1976). At the consultation meeting on this programme, held in San José on 3-4 May 1978, a list of priority activities was drawn up. It is interesting to note that first priority was assigned to a renewable source of energy—the project for manufacturing alcohol to be used as a fuel. Also among these priorities was a project for the use of agricultural and forest wastes for generating energy.

In almost all the countries visited, organizations exist with functions related to the production or the consumption of energy. There are electric power companies responsible for the development of electrical systems and mining, energy and hydrocarbons departments or ministries responsible for exploring and exploiting oil or natural gas deposits and for the commercialization and distribution of fuels. In addition, there are transportation and industry

---

\* Decree No. 6066 of the Ministry of Economy, Industry and Commerce, published in the Official Gazette, July 2, 1976. This committee was composed of two representatives of the Ministry of Economy, Industry and Commerce; one representative of the Ministry of the Presidency; one representative of the Ministry of Public Works and Transportation; one representative of the National Electricity Service; one representative of the Costa Rican Development Corporation; one representative of the Costa Rican Oil Refining Company and one representative of the private sector.

departments or ministries that supervise activities that consume fuel and, lastly, organizations that are concerned with agricultural and rural developments, for which an adequate supply of energy is important. The multiplicity of organizations engaged in one aspect or another of the energy problem makes it essential to establish some kind of institutional machinery at the national level to co-ordinate the activities of such institutions and to assist in formulating energy policy criteria that consider the problem in its totality.

## INVENTORY OF RESOURCES OF RENEWABLE SOURCES OF ENERGY

ANALYSIS of the extent of knowledge of the resources existing in the various countries included in this study points to significant differences in the amount of information on hand, both with respect to various sources of energy and among countries with respect to a single source of energy. Theoretically, various institutions are responsible for collecting information on the availability of resources, such as meteorological services or institutes, universities, air forces and agricultural research institutes. Naturally, these institutions are usually specialized and centre their attention on only one or two sources of energy. From the information gathered by the present study, one may observe that some countries possess practically no information at all nor do they possess institutional machinery for obtaining it; others have been able to gather sufficient data to determine which sources of renewable energy might be used in specific regions.

The problem of gathering information about the availability of energy resources implies both the need for institutions to carry out this task and a sufficient force of well-trained personnel supplied with the required equipment and instruments to ensure accuracy of assessment. It is also necessary for these or other institutions to systematize and process the information collected; otherwise, the situation observed during some of the visits made for this study may be repeated: assessment data stored in files where they serve no useful purpose have accumulated since no one has taken the trouble to process such information.

The role renewable and non-commercial energy has been playing in the supply of energy is also worth considering. Energy inventories normally have included only commercial energy in such inventories, even when a substantial portion of the energy consumed by the population is supplied non-commercially.* Only a few countries have incorporated this source in their energy inventories. Such is the case in Brazil where, for a few years now, non-commercial energy has been included in its inventory; in Uruguay, where wind energy is included; and in Peru, where a United Nations project is presently underway to improve the methodology for formulating energy inventories, including non-commercial energy.

* Approximately 40 per cent of the energy consumed in Central America is non-commercial energy, according to information provided by SIECA.

# SCIENCE, TECHNOLOGY AND INSTITUTIONAL MACHINERY

INSTITUTIONAL MACHINERY to provide support to the scientific and technological sector may be of great importance in the development of renewable sources of energy. Consequently, detailed knowledge of the technologies for using these sources is an essential prerequisite either for transferring the most appropriate technologies to local realities or for developing new technologies. The problems involved in scientific and technological development are intimately linked with general problems of development and are relatively similar in any given field of research, such as housing, agriculture or energy. Generally speaking, one may say that analysis of scientific and technological systems may be carried out by considering three subsystems: information, education and transfer of technology.

## Information

One of the recurring topics in the interviews with researchers in the field of renewable sources of energy was the need for more information on research being performed on this subject both within the region and in the industrialized countries. At present, such information may be obtained from scientific publications or by attending congresses or seminars on this topic; however, in both cases the information is obtained with a certain delay that prevents researchers from using it to improve the orientation of their own research.

Certain private organizations exist, such as the Latin American Solar Energy Association (ALES), which enable those working in this field to maintain contact with one another; nevertheless, such organizations have the disadvantage of covering only a partial aspect of the entire problem.

## Education

The *sine qua non* condition for the success of any line of research is the existence of a 'critical mass' of researchers working in a single area. One of the first tasks to be carried out by countries interested in the development of renewable sources of energy is to form a basic nucleus of people specialized in the various sources involved. Scientists must be encouraged by appropriate incentives to choose renewable sources of energy as their main interest area. In view of the diverse aspects involved in the development and introduction of such sources of energy into the rural sector, which demand consideration of the social, cultural and economic characteristics of the potential users, it is necessary to organize interdisciplinary research groups that take these factors into account.

It is extremely important for the educational subsystem to create awareness among primary and secondary school students of the energy problem that humanity will soon face and of the importance of making economical use of non-renewable resources and promoting the development of renewable sources of energy. A generation of young people educated to be aware of the energy problem may possess values and attitudes different from those of the present generation, in which large and cheap supplies of energy have led to much waste and squandering of resources.

**Transfer of Technology**

One of the factors that has contributed to the inability of developing countries to respond appropriately to the challenge of technology has been the conditions under which the technology developed in the industrialized countries has been transferred to them. In recent years, thanks to the efforts of UNCTAD and other United Nations organizations, much greater awareness has been created with respect to the problems associated with the transfer of technology, such as the cost of such transfer, the various clauses that are generally included in transfer contracts and that prevent the developing countries from obtaining significant benefits and the negative long-term effects resulting from the purchase of inappropriate technologies and the ensuing inhibition of national technological development. These problems will emerge again in the transfer of technology to the use of renewable sources of energy if steps are not taken to develop national technological capacity.

    The problem is all the more serious in view of the large amounts of resources the industrialized countries have been allocating for some time to the development of renewable sources of energy, such as solar energy programmes or projects for constructing high-powered windmills to be hooked-up to power grids.

It may be concluded that two parallel and complementary energy markets exist: firstly, a market for sources of energy such as oil, gas, coal and uranium, and secondly, a market for energy technologies.* Some underdeveloped countries export energy resources; nonetheless, they are usually highly dependent on technologies. The market for energy technologies is a closed market with formidable barriers to prevent the entry of new firms, a fact that places the underdeveloped countries in a weak negotiating position. This may be observed in the market for conventional energy technologies where practically no autonomous technological development has taken place in the underdeveloped countries, which have simply copied already developed technologies to generate electricity in large power stations. An example of this is the case of thermal generation of electricity, which has been known in the region for more than one hundred years and has always employed imported technology, practically without adaptation. However, some Latin American countries have created the technical capacity necessary to construct these kinds of power stations, especially water-powered stations. In this area, *Endesa Chilena* and the Brazilian electrical sector have made

---

\* This idea has been developed in an UNCTAD report on the transfer of technology in the field of energy.

significant advances in local engineering and Brazilian companies today provide advisory services on water projects in Africa and the Middle East.

In order to minimize the problems associated with the transfer of technology using renewable sources of energy, various Latin American countries have begun to establish instruments for controlling such transfers. In this respect, it is interesting to note the Brazilian experience in this area and Mexico's National Registry of the Transfer of Technology, which is empowered to review all contracts for the transfer of technology subscribed to by Mexican entrepreneurs with foreign companies. So far, these countries have not registered any transfer contracts for equipment using renewable sources of energy, an indication that the same is probably true for the entire region.

# MANUFACTURE, MARKETING AND DISTRIBUTION OF EQUIPMENT USING RENEWABLE SOURCES OF ENERGY

THE DEVELOPMENT of renewable sources of energy in Latin America demands that its countries be able to design and manufacture equipment to transform such sources into thermal, mechanical or electrical energy. The industrial capacity to produce such equipment exists in most of the countries in the region, albeit in varying degrees. Obviously, it will require much greater effort to increase quality as well as reduce costs, which in reality means increasing the capacity of local engineering to improve the design of this equipment. If this is not accomplished within a few years the region will see the introduction of mass-produced equipment in the industrialized countries, probably in kit form.

In certain cases — for example, with regard to solar equipment — the problem consists not so much of a lack of knowledge about solar technology *per se*, as of improving materials technology to increase the useful life of such equipment. Thus, although capacity exists in these countries, an enormous amount of ground still remains to be covered in order to perfect equipment using renewable sources of energy. This problem has obvious institutional implications insofar as it will be necessary to secure much closer collaboration between researchers and the universities on the one hand, and the manufacturers of equipment on the other, in order to solve existing design and materials problems jointly.

The existence of factories for the manufacture of equipment using renewable sources of energy in large urban centres does not ensure its distribution and marketing in rural sectors, precisely the area in which such equipment is most necessary. In the case of almost any commercial product, demand normally becomes the incentive for setting up distribution points throughout the national territory, without the need for institutional support to create such a network. However, the situation is different in this case, in which it would be in the interest of the State to publicize and promote the installation of equipment using renewable sources of energy, since the advantages of their use are perhaps not fully realized by potential users. This situation makes closer collaboration necessary between government organizations responsible for rural development and manufacturers or researchers desiring greater distribution of the equipment they are producing, whether on a commercial or experimental level. If it is hoped to promote greater use of this equipment, a marketing strategy must be formulated either by using already existing channels, such as agricultural banks and extension services, or by creating entirely new channels specifically for this purpose.

Another aspect related to distribution is the legal situation prevailing in a given country with respect to the use of some of the sources of energy discussed in this study. For example, in Brazil, the use of small waterfalls is controlled by the government; farmers wishing to use them must request a concession. Limitations of this kind may also exist in other countries.

## Means of Communication and Responsible Institutions

An essential factor in the capacity of countries for developing renewable sources of energy is the belief by all sectors of the population — especially rural sectors — in the prospects and advantages offered by such sources. This requires intensive efforts in which both public and private means of communication should play an important role. For this purpose, high-level organization of a commission responsible for designing and implementing a publicity campaign on the advantages of renewable sources of energy may prove to be effective. Researchers working in this field should strive to translate the results of their efforts into ordinary language easily understood by people with little education so that the advantages and savings that may be obtained through the use of such equipment will be readily grasped. Certain organizations already existing in almost all countries, such as agricultural extension services with experience in introducing new agricultural technologies, could be instrumental in carrying out this activity. The development of pilot projects as demonstrations may also help to awaken interest in this equipment.

## Institutional Machinery and Financing

All the institutional mechanisms mentioned require substantial financial support. Consequently, all the elements constituting the chain of development of equipment using renewable sources of energy — researchers, manufacturers, distributors and users — would benefit from access to financing that would assist them in researching, manufacturing, selling or buying this kind of equipment. The State should investigate special financing methods and mechanisms to develop this equipment since, although private investment in its development may not be highly advantageous, the social benefits to be derived are quite evident. Also, the principal problem involved in installing such equipment may be more a financial than an economic one since, although the savings achieved may be great, the users may not possess funds to make the required investment.

The various problems outlined here demonstrate that development of renewable sources of energy requires creating or improving diverse institutional machinery, so that formulating and implementing the policies required to achieve this objective will be more feasible.

## References

National Commission for Saving Energy (1976) *Report.* San José, January.

UNDP (1976) Project Document N. RLA/76/012, p. 3.

CHAPTER 7

# CASE STUDIES

## I Small Hydroelectric Plants in Colombia

*Ing. Mercy Blanco de Monton*
Instituto Colombiano de Energia Electrica

THE GOVERNMENT of Colombia, through the Ministry of Mines and Energy and its organisation, the Colombian Institute of Electrical Energy (ICEL), is in charge of developing those numerous populated zones that presently are isolated from the developed centres due to their geographic location. Because of distance or difficult topography, these towns have been unable—from a technical and economic point of view—to get the benefits of energy through thermal transmission or generation as a short-term solution. Therefore, the ICEL has advised these towns to make use of nearby water sources for hydraulic energy. To make this solution a reality is the objective of the small hydroelectric plants, whose methodology and initial performance will now be discussed.

### Location of the Plants

The Colombian territory is divided politically into *departamentos, intendencias* and *comisarias.* The ICEL acts through the electrical enterprises that have been organized in each *departmento,* such enterprises undertaking the generation, transmission and administration programmes in the areas assigned to them. So far, these enterprises have identified those zones left out of the energy plans; moreover, they have also obtained a great deal of the technical, economic and social information which has helped in the definite selection of the study centres.

### Scope of the Plan

Thirty-five places have been selected, with the following priorities in mind: (1) places with more information in relation to designs of restitution and flows; (2) places with restitution designs but without available information on flows; and (3) places with information on flows but without available restitution design.

## Characteristics of the Plan

In elaborating on the studies' plan for the construction of small hydroelectric plants, researchers established the following steps:

(1) The studies will be made with Colombian personnel only.
(2) The equipment and elements to be used during the construction of the small hydroelectric plants will be national, as far as possible.
(3) In general, the places that have water streams with a high average flow will be considered, since it is possible that the water currents to be used will not have historic registers or, if they do, the registers may not be accurate. In these cases, installed potential will be calculated on the basis of minimum flows.
(4) Equipment will be standardized as much as possible to decrease maintenance problems and to facilitate transport and assembly.
(5) Materials available in the region should be used as much as possible to avoid an overcharge in transportation costs.

## Execution of the Plan

The corresponding studies of the small hydroelectric plants have been divided into three stages: (1) preliminary recognition (identification); (2) feasibility study; and (3) design of the best utilization, including plans of bids for construction.

In the most simple projects, stages 2 and 3 will be considered as one. Many national consultant companies were invited to carry out the plan; five were chosen to proceed with the studies, with a total of seven sites per company. The studies were initiated in the last quarter of 1978 and were finished by mid-1979, so that the opening bids for the construction of the works could begin immediately. In the first quarter of 1978 Stage 1 was completed and studies corresponding to Stage 2 were begun.

## Methodology Used

The different parameters that establish the characteristics of hydroelectric utilization include the following:

### (1) Calculation of the Energy Demand

The analysis of the energy demand in the studied zone was done with the usual methodology, within the following criteria:

(a) *Growing rate of population:* Values between 1.5 and 3.5 per cent per year were used according to the economic potential of the region;
(b) *Projection of the Demand:* A period of 15 years was considered convenient.

(c) *Energy Consumption:* Based on the economic balance of the regions to receive benefits, the consumption value adopted varied in the following way:

    consumption per inhabitant : 0.110 - 0.210 kW
    consumption per household : 0.6 - 1.5 kW
    commercial consumption : according to present and future characteristics of the region;
    industrial consumption : according to economic potential and developing facilities of the region;

(d) *Growing Rate of Demand:* Around 3 per cent annually.

## (2) Topographic Studies

(a) *Recognition Stage.* The levels of inlets, loading chamber and workshop were established. The conduction line and possible route of the transmission line were defined.

(b) *Feasibility Stage.* Detailed topographic surveys were done at a level that would permit elaboration of the designs and pre-designs of the civil works and evaluation of the magnitude of movements of the land required. The structure plans were elaborated in scales 1/200-1/500.

(c) *Design Stage.* Topographic studies consisted in the location of the definitive studies and eventually, if necessary, in more detail of the interest sites.

## (3) Hydrological Studies

(a) *Recognition Stage.* Existing information was analyzed, defining the scope of its utilization and accuracy. The usefulness of the present hydro-meteorological network was established, as well as its possible complementation directed towards the efficiency of the project. Activities at this level of the studies were directed towards obtaining the value of minimum flows, using conventional methods at a preliminary level. For this purpose, the water sources to be used were gauged in order to define the magnitude of the transported flow especially during low-water periods. Information was gathered from the inhabitants of the region in relation to water levels reached by the crescents and their occurrence. The following procedures were included:

(a) evaluation of the topographic characteristics—geomorphological and hydrological—of the studied zone;
(b) estimation of the minimum flow as a medium percentage flow (A MIN = 2/3 Q gauged);
(c) on the sites where cartography details existed, the area of the basin and its hydrological readiness were determined;
(d) comparative analysis of rain-duration curves;
(e) flow-duration curves;
(f) probable function of minimum flows.

The importance of the value of the minimum flow was given by the hydroelectric utilization for those flows, which preferably should be at water level.

(b) *Feasibility Stage.* The studies corresponding to this level could be summarized, in general terms, in the execution of the required studies for a most reliable support for the values of minimum flow, obtained during the recognition stage, at a grade that will allow starting the designs of the works. As general activities performed during this stage, the following could be considered:
- (a) installation and management of hydro-meteorological stations programmed during the recognition stage;
- (b) gauges of the water source to be used;
- (c) analysis and interpretation of the hydro-meteorological data and correlation studies;
- (d) sediment analysis.

(c) *Design Stage.* During this stage, operation of the hydro-meteorological stations has continued, so that continuity of the data will not be lost.

**(4) Geological Studies**
(a) *Recognition Stage.* Maps of general geology were elaborated which will enable establishment, in a preliminary form, of the geological conditions of the zones where the utilizations have been located.
(b) *Feasibility Stage.* Geological and geotechnical studies (semi-detailed) were performed. The geotechnical studies include the execution of exploration and trenches to determine the mechanical characteristics of the soil. Also, geomorphological studies to determine the areas containing sediments were performed and the sources of construction materials were identified and evaluated.

**(5) Ecological Studies**
(a) *Recognition Stage.* The different physiographic aspects were studied initially, as well as the climatological, agricultural and forestry soils and the fauna.
(b) *Feasibility Stage.* The results detected in the first stage were confirmed.

**Types of Utilization of this plan**

The characteristics of the utilization found in the first stage of the plan (Small Hydroelectric Plants) allow for two types of development, according to the magnitude of their fall: (1) utilization of medium fall, with H values going from 30 m and 120 m; and (2) utilization of the flow, with H values less than 5 m.

## Cost Analysis

Within the studies performed for the plan 'Small Hydroelectric Plants', corresponding costs for each of them in terms of a total cost were established; the costs were separated into local currency and foreign currency, as well as the index costs in US dollars per kilowatt. The costs take into account investments of civil works, electro-mechanical equipment and transmission lines. Unforeseen costs were not taken into consideration—which for these types of project vary from 15 to 20 per cent—nor were financing costs during construction.

To evaluate the economic benefits of the small hydroelectric plants, as against the alternatives of diesel-generated plants, comparative studies for all the projects were performed. From these, we are appending to this case study the analysis for utilization in the Choco region, where—and it is worthwhile noting—the most difficult conditions are found, due to difficult access of the locations, extreme rains and precarious socio-economic conditions.

Table 7.1 summarizes the cost of civil works, electro-mechanical equipment and transmission lines to the populated centres, as well as the total cost of the project and the unitary cost in US dollars per kilowatt for the recommended alternatives to each of the following projects: Unguia, Nuqui, Jurado and Bahia Solano.

Figure 7.1 compares the cost curves of thermal diesel generation and hydraulic generation in US dollars per kilowatt for the specific recommended projects in Unguia, Bahia Solano, Jurado and Nuqui. These curves are defined in terms of cost in US dollars per kilowatt, pre-set value *versus* the return rate from 2 per cent up to 18 per cent. In the same figure, two family curves of thermal generation are presented, one with a fuel cost of $40 per gallon with potentials of 100 to 1,000 kW and the other one, with a fuel cost of $50 per gallon and potentials of 100 to 1,000 kW. The costs of personnel for diesel generation and for hydraulic generation have been evaluated in the first approximation for the economic calculation performed.

## RESULTS FROM THE FIRST STAGE

THE RESULTS obtained in the first stage of preliminary recognition have been quite satisfactory in terms of the technical level of the corresponding reports and the number of suitable sites that have been found and that deserve further study to the feasibility and design stages.

The fulfilment of activities in the preliminary recognition stage had the following results:

(1) Twenty-seven of the chosen sites justified the continuation of the studies.

Table 7.1  Total and Unitary Costs of the Project

| | Unguia | | | | Nuqui | | Jurado | | Bahia Solano | |
|---|---|---|---|---|---|---|---|---|---|---|
| | Albania (650 kW) | | Cuti B (325 kW) | | Sitio II (700 kw) | | Sitio II B (750 kW) | | Mutata 1a (1,100 kW) Etapa | |
| Description | Pesos | US $ | Pesos | US $ | Pesos | US $ | Pesos | US $ | Pesos | US $ |
| Civil Works Subtotal* Cost/kW | 20.047 40.995 | 642 988 | 18.276 56.234 | 440 1.355 | 17.155 24.507 | 413 591 | 17.471 23.295 | 421 561 | 27.281 24.801 | 657 598 |
| Electro-mechanical Equipment Subtotal* Cost/kW | 32.675 33.346 | 522 801 | 10.964 33.735 | 204 813 | 18.289 26.127 | 441 130 | 22.739 30.319 | 548 731 | 17.391 15.810 | 419 381 |
| Transmission Lines Subtotal* Cost/kW | 4.926 7.578 | 119 183 | – – | – – | 4.876 6.966 | 117 168 | 4.110 5.480 | 99 132 | 3.502 3.184 | 84 77 |
| Project Costs Total Cost per kW | 53.248 81.920 | 1.283 1.974 | 29.210 89.969 | 705 2.168 | 40.320 57.600 | 971 1.388 | 44.320 59.094 | 1.068 1.424 | 48.174 43.795 | 1.161 1.055 |

*In thousands

# CASE STUDIES: SMALL HYDROELECTRICITY IN COLOMBIA

**Figure 7.1** Thermo-Diesel Generation Cost and Hydroelectric Generation Cost (from the Unguia, Jurado, Nuqui and Bahia Solano Projects) versus Rate of Return

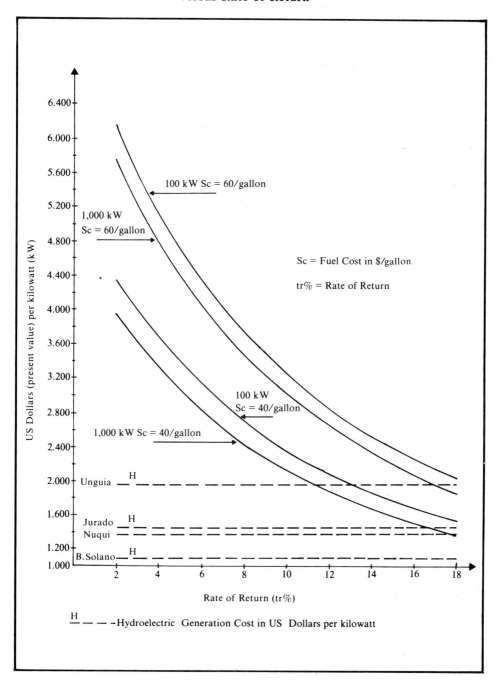

(2) Although their development could involve high investment costs, those sites whose special border situation gave them unquestionable importance in the defence of the national and territorial integrity of the country, were given priority.

(3) The energy demand of the studied populations, projected for 15 years, was over 5,000 kW in a few cases.

(4) In general, studies of certain sites were suspended for the following reasons:
   (a) excessive cost of generated energy;
   (b) lack of hydroelectric energy sources nearby;
   (c) sources with possibilities of hydroelectric utilization near populations, but of a larger magnitude than the one established for the small hydroelectric plants;
   (d) the possibility of energy supply through the transmission lines.

It is important to emphasize the identification, within the studied zones, of relatively important population centres whose energy supply—although not considered in the initial plan — could be obtained with the installed capacity of the utilizations under study, thereby increasing the social and economic benefits of small hydroelectric plants. Such circumstances considerably raise the number of population centres that would be favoured and compensate somewhat for the ones that had to be eliminated.

## CONCLUSIONS

(1) The benefits of the plan cannot be denied, in each of the forms considered since (notwithstanding, from an enterprise point of view, the scarce or null economic retentiveness to justify this plan) it is expected that the Keynesian Multiplier Effect will operate in this type of investment, if one takes into consideration that energy is basic to development.

(2) The population centres chosen as the object of this study constitute a statistical sample on the potential of specific resources, fully identified in mining, agriculture, cattle-breeding, fishing, forestry and tourism.

(3) The people of the centres studied live under precarious conditions and, with few exceptions, lack the services of physical and social infrastructures.

(4) Installation of small hydroelectric plants makes it possible for the people of these regions to develop small industries and crafts that will contribute to improving living conditions by providing new work sources.

(5) Preliminary economic analysis demonstrates that the developing hydroelectric plants presently under study are avoiding the alternatives of thermal generation, which are more costly and little trusted, due to the difficulties of having a continuous supply of fuel and maintenance.

# II  Development of Biogas Plants in Guatemala (1952-1979)

*Mario David Penagos Gozalbo*
OLADE Consultant

## RESEARCH AND EXPERIMENTATION (1952-1958)

IN 1952, while still a student at the Faculty of Agronomy in the University of San Carlos de Borromeo of Guatemala, I attended a lecture by engineer Franz Billeb Vela. The lecturer mentioned, among other experiences, the studies made by two eminent French engineers in the laboratories of the School of Agriculture of Algeria. The objective was to produce, by the most rapid and suitable means, the decomposition of organic matter. It was the first experiment on the anaerobic treatment of material taken from agricultural media and new benefits were obtained through the use of this procedure. A most important result was the production of a gas, formed during fermentation, which could be used as a fuel.

The enthusiasm associated with this research was transformed into action and a small *in vitro* installation was prepared for testing. Different fermentations were made with several samples of horse manure. The first sample to give off gas was introduced into the small installation. During the next few days, enough gas had accumulated to conduct tests with a Bunsen burner. This initial success encouraged us to build a small metallic plant comprising two fermentation chambers, each different, in order to assess which would give the better result with gas pressure already connected to the gasometer. During the next six months, this miniature plant enabled us to compile data for use in a larger model. New waste materials were tested: chopped corn stalks and wheat chaff. The problems of hydraulics, gas flow, pressures and calculations were known and tested.

In 1953, the construction of a pilot plant was begun, consisting of two fermentation chambers and a gasometer. With the pilot plant in operation, many tests and demonstrations were made on burners, gas burners, regular propane gas ranges, gas burning lamps and on several vehicles, which were made to run. All the tests were satisfactory and demonstrated the high hydro-carbonic power of the biologic gas. After three months' operation, the plant was emptied and a substantial amount of fertilizer was obtained for use and analysis. Analyses were made in several research centres: reports on the fertilizer content and its effects on the physiology of the plant (through a change in texture and structure of the soil) placed it advantageously among the conventional manufacturing systems of 'humus'.

The gas was subjected to analysis by the Institute Centro-americano de Investigacion y Technologia Industrial (ICAITI), which reported a mixture of 68 per cent methane, 31 per cent carbon dioxide, 1 per cent oxygen and traces of other gases. In 1954, demonstrations were made to various interested groups: farmers, livestock breeders, industrialists, research institutions and government authorities.

In 1956, former President Colonel Carlos Castillo Armas was greatly impressed with a demonstration of the functioning of a stove, an electric solderer and an automobile. He immediately authorized the acquisition of essential research equipment through the Ministry of National Defense. He also authorized a report on bibliographic research from ICAITI, together with instruments and machinery for *ad hoc* quotations on equipment for the purpose of building a plant of greater capacity. However, his tragic death in 1957 prevented the operation of this decisive project. Regarding the requested report, the ICAITI (1956) stated: 'Although the work done went beyond a simple industrial consultation, as defined in Resolution CD/NI/56/RI of the Board of Directors of this Institute, and consequently should be subject to the charges for costs incurred, ICAITI is pleased to submit the attached report without charge, in view of the project's potential importance to Guatemala. Signed: Alberto Mides, Director, ICAITI.'

In 1957, former President Colonel Guillermo Flores Avendano requested a demonstration at the pilot plant; he was most interested in having the procedure publicized and offered his full support to put it into operation. He was informed that, in Europe, work was being carried out on the same project and that it would be convenient to know of the progress achieved before proceeding to the promotion considered. As a consequence, he granted three scholarships—two through the Ministry of Economy and the other through the Institute for the Development of Production (INFOP). Engineer Franz Billeb Vela and technician Rafael Penagos Gozalbo were invited to accompany the author to Europe in 1958 to study developments in this field.

The experience was extremely interesting, especially in Germany where large plants for the treatment of sewage water were operating and US$ 180 million were being invested in the promotion of rural constructions. The systems used were highly technical (such as the Allerhop, Schmidteggersgluss and Darmstadt), with experiments being conducted in Germany, France, Italy and England. In our trip report, we mentioned that this technology was little suited to our medium not only because of the control required for its operation but also because of the skilled labour needed for its installation and the high cost of construction. The Guatemalan system was the most appropriate for Latin America because of its simplicity, easy management and low cost.

In 1958, the incumbent President of the Republic, General and Engineer Miguel Idigoras Fuentes, attended a new demonstration at the pilot plant, where he observed the operation of a gas range, a generator group and an automobile in which no change was made in its carburation system. He said, at that time, 'I am a sponsor of this procedure. Four or five of these plants should be built on the outskirts of the Capital so that everyone can see them.' In spite of his good intentions, regarding easy loans to be granted by INFOP to parties interested in building a plant, nothing came of them since the institution did not have the funds.

## PROMOTION AND DEVELOPMENT OF ENTERPRISES (1958 - 1979)

IN 1958, an enterprise was created called Abiogasco Fertilizers and Biologic Gas Company, for the purpose of promoting and developing the installation of biologic plants in rural areas of Guatemala. The advantages to be obtained for the country from the use of the system and its products were evident. The procedure was patented and an advertising campaign conducted. More than 200 demonstrations were made to the press, agricultural and livestock institutions, interested individuals and banking circles. All expressed their admiration and wishes for success. Nevertheless, the resultant anecdote is warranted: a government minister attended one of the many demonstrations and after expressing his admiration and good wishes, upon leaving he said to someone accompanying him: 'These people think they have fooled us; inside those battery cells they must have a propane gas cylinder.' His companion had helped us in filling and emptying the chambers more than once.

In 1958 and 1959, front-page headlines were published as follows:

*Demonstrations with Biologic Gas were Successful.* Perspectives for producing electric power and fertilizers. Tests conducted by Engineer Penagos in presence of Minister of Agriculture; remarkable results after seven years of studies. *El Imparcial,* 14 July 1958, No. 12080.

*Invention for benefit of Farmers. Economic Pulse. Prensa Libre,* 12 August 1958.

*Successful Experiment: Gas produces Automotive Power.* Two Guatemalans successful in producing energy from biologic gas. Effective substitute for gasoline, diesel, etc. Tons of organic fertilizer may also be produced as byproduct. *Prensa Libre,* 7 August 1958, No. 2175.

*Industry of Great Potential to be started soon in Guatemala.* Production of biologic gases and organic fertilizers using raw materials existing in abundance in our country. *La Hora,* 14 July 1959, No. 4821.

The following installations were then constructed in Guatemala (Table 7.2). In 1959, Don Justo Abascal de Anda, a farmer, entrusted the construction of the first biologic plant to us, Type Guatemala 10, with the following features: steel construction, two fermenting chambers of 10 m$^3$ each and a gasometer of 10 m$^3$ with steel vat and chamber. The apparatus was portable, enabling transportation to any of his farms. The Banco del Agro was the first institution to offer and grant, without cumbersome negotiations, credit for this purpose. In 1960, Mr Abascal's plant was exhibited at the First Spring Fair held in Guatemala City and attracted

**Table 7.2  Construction of Plants producing Fertilizers and Biologic Gas in a single process**

| STUDY AND DESIGN | | | | EXECUTION | | | | | |
|---|---|---|---|---|---|---|---|---|---|
| Name, Plant Owner Location | Plant Type, Material, Construction | Time of Construction, Beginning, End | Temperature Climate-Site (+/-°C) | Raw Mat. Class Quantity (quintals) | Capacity Production Biofertilizer* (year quintals) | Biogas** At. Pres. (year m³) | Total amount invest. (quetzals)*** | Value of Products Biofert. (quintals + biogas m³) | Period of Amort. Investment (years) |
| *Pilot Plant* Estrada Penagos Guatemala City 4 a.c. & 2a.Av.Z.10 | G-10 Pilot brick | 8 months V/1953 XII/53 | Temperate semidry climate 22°C | Various Agricultural 1,320 | — 1,200 | — 2,400 | 2,500.00 — | 3,600.00 | 1 |
| *Chipo* Justo Abescal Santa Barbara Escuintla | G-10 Portable rolled steel | 3 months VII/59 IX/59 | Warm Humid climate 30°C | Cattle Manure 1,320 | — 1,200 | — 2,400 | 3,000.00 — | 3,600.00 | 2 |
| *San Alberto* Edmundo Vasques Patulul Escuintla | G-15 Farm reinforced concrete | 5 months III/59 VII/59 | Non-condt. Warm Humid 28°C | Cattle Manure coffee-pulp 1,650 | — 1,500 | — 3,000 | 5,000.00 — | 4,500.00 | 2 |
| *Municipal Plant* Guatemala City 21 C. & 6a.Av.z.1 | G-120 Sanitary reinforced concrete | XI/59 XII/60 | Non-condt. Temperate semidry 22°C | Market Garbage 22,000 | — 20,000 | — 40,000 | — — | 24,000.00 | 2 |
| *El Porvenir* Antonio Bonifasi Mazatenango Suchitepequez | G-90 Agro-Ind. reinforced concrete | 12 months VIII/59 VII/60 | Condt. Warm Humid 30°C | Coffee-pulp Slaught. wastes 16,500 min. | — 15,000 Minimum | — 30,000 Minimum | 35,000.00 — | 18,000.00 | 3 |
| *Colima* Jose Bonifasti Zunilito Suchitepequez | G-45 Agricultur. reinforced concrete | 6 months XII/61 V/62 | Non-condt. Temperate Very humid 24°C | Coffee-pulp Stubble 6,600 | — 6,000 | — 12,000 | 12,000.00 — | 7,200.00 | 2 |

*continued*

# CASE STUDIES: BIOGAS PLANTS IN GUATEMALA

| STUDY AND DESIGN | | | | | EXECUTION | | | | | |
|---|---|---|---|---|---|---|---|---|---|---|
| Name, Plant Owner Location | Plant Type, Material, Construction | Time of Construction, Beginning, End | Temperature Climate-Site (+/-°C) | Raw Mat. Class Quantity (quintals) | Capacity Production | | | Total amount invest. (quetzals)*** | Value of Products Biofert. (quintals + biogas m³) | Period of Amort. Investment (years) |
| | | | | | Biofertilizer* (year quintals) | Biogas** At. Pres. (year m³) | | | | |
| *La Esperanza* Ricardo Echeverria Pochuta Chimaltenango | G-45 Agricultur. reinforced concrete | 6 months XI/62 V/63 | Non-condt. Temperate Humid 24°C | Coffee-pulp Stubble | — 6,000 | — 12,000 | | 12,000.00 | 7,200.00 | 2 |
| *La Esperanza (expansion)* Ricardo Echeverria Pochuta Chimaltenango | G-30 Agricultur. reinforced concrete | 5 months VII/63 XII/63 | Condt. Temperate Humid 21°C | Coffee-pulp 5,500 | — 5,000 | — 10,000 | | 5,000.00 | 6,000.00 | 1 |
| *El Milagro* Ricardo Remmele Granja El Trebol Co. El Milagro | G-75 Cattle Sanitary reinforced concrete | 4 months XII/64 VI/65 | Non-condt. Temperate Dry 23°C | Cattle and pig manure 15,200 | — 12,000 | — 24,000 | | 15,000.00 — — | 14,400.00 | 2 |
| *La Sierra* Rodolfo Coutillo Patzun Chimaltenango | G-60 Cattle reinforced concrete | 4 months II/66 VI/66 | Non-condt. Cold Humid 18°C | Cattle manure 11,000 | — 10,000 | — — | | 8,000.00 | 10,000.00 | 1 |
| *Biofert* Alvarado-Lozano Los Sauces Pali, Escuintla | G-300 Cattle-Industrial reinforced concrete | 7 months V/69 XII/69 | Non-condt. Temperate Humid 24°C | Manure Various 55,000 | — 50,000 | — — | | 80,000.00 | 25,000.00 (02.50/qq) | 2 |
| *Rancho Alegre* Rosendo Gordillo Rio Bravo Escuintla | G-75 Cattle reinforced concrete | 6 months VII/69 I/70 | Condt. Warm Dry 28°C | Cattle manure 17,600 | — 16,000 | — — | | 22,000.00 | 16,000.00 | 2 |

*continued*

| STUDY AND DESIGN | | | | EXECUTION | | | | | |
|---|---|---|---|---|---|---|---|---|---|
| Name, Plant Owner Location | Plant Type, Material, Construction | Time of Construction, Beginning, End | Temperature Climate-Site (+/−°C) | Raw Mat. Class Quantity (quintals) | Capacity Production | | Total amount invest. (quetzals)*** | Value of Products Biofert. (quintals + biogas m³) | Period of Amort. Investment (years) |
| | | | | | Biofertilizer* (year quintals) | Biogas** At. Pres. (year m³) | | | |
| *Las Flores* Fernando Luna San Antonio Suchitepequez | G-60 Cattle reinforced concrete | 6 months VIII/69 II/70 | Condt. Warm Dry 28°C | Cattle manure 11,000 | 10,000 | — — | 8,000.00 | 10,000.00 | 1 |
| *Las Acacias* Milton Molina E. La Gomera Escuintla | G-200 Pilot reinforced concrete | 18 months VII/77 XII/78 | Non-condt. Warm Dry 30°C | Cattle manure Slaught. wastes 32,000 | — 28,800 | — 73,000 | 46,100.00 | 36,100.00 | 2 |

\* 1 quintal (qq) = 100 English pounds with 30 per cent humidity and a minimum value of US$1.
\*\* 1 cubic meter (m³) at atmospheric pressure; 6,000 Kcal, which equals 23,800 BTU.
\*\*\* Quetzal = US$1.
Slaught. wastes = Slaughterhouse wastes.

much interest. In this same plant, a laboratory test was conducted for the first time to obtain biofertilizer and biogas from sugar cane bagasse. The test was successful — the yield of gas was equal to $0.6/m^3$, with an average atmospheric temperature of 22°C. The gas burned perfectly and the biofertilizer yield had 15 per cent less organic matter content than that used as raw material and which had been transformed into gas. In field tests, at higher atmospheric temperatures averaging 30°C, the gas yield was equal to $1.2/m^3$.

These demonstrations and exhibition of an electric generator, a gas range and a gas refrigerator in operation in a pre-fabricated house, in addition to the publicity gained, resulted in the construction of the following plants.

In 1959, Edmundo Vasquez requested the construction of a small Hacienda-type, G-15, on his farm *San Alberto* in Escuintla. This stationary plant used cow manure and humid coffee pulp remains and, within five months, was allowing a generator motor of 3.5 kWh to be put in motion to light the house (farm). The energy was used to operate an electric pump for a well which supplied the necessary water; pipes carried gas to the kitchen, where it was used for the stove or refrigerator.

In 1959, the farmer Antonio Bonifasi requested that we prepare three industrial-type projects for his farm *El Porvenir* in Suchitepequez, to take advantage of the garbage of Mazatenango city. The proposed plants were:

(a) Industrial Type, for using the domestic waste of Mazetenango City;
(b) Agro-industrial Type, using pulp remains of 3,000 quintals (1 quintal is equivalent to approximately 46 kilograms) of coffee *oro,* in addition to slaughterhouse wastes;
(c) Agricula Type, for preparing compost from pulp remains of coffee (without use of gas).

Of these, the second type was chosen; its construction lasted 12 months and operations started in July 1960.

In 1959, the Mayor of Guatemala City, Doctor Luis F. Galich, requested us to plan a plant to treat the garbage of a market located in the civic centre of the city, at a distance of 100 metres from the Town Hall. The work was contracted under our direction and supervision. But materials, equipment, instruments and labour were provided by the municipality. In 1960, when 75 per cent of the engineering work had been completed, the land was transferred to the Ministry of Public Works who had reserved it for other purposes and the municipal biologic plant was demolished.

In 1960, José Bonifasi, whose brother owned the *El Porvenir* plant, had seen the results obtained and requested us to construct a plant on his farm Colima in Suchitepequez. Construction was begun in 1962 and completed in six months. The plant's function was to process pulp obtained from 2,000 quintals of *oro* coffee, together with the waste from ten cows.

In 1962, Ricardo Echeverria Herrera requested us to build an Abiogasco plant on his farm *La Esperanza* in Fochuta Municipality, Chinaltenango Province. The plant would partially process the remaining pulp of 3,000 quintals of *oro* coffee and the gas would be used to dry the coffee. Construction of the plant was completed in June 1963. In 1963, Echeverria requested the enlargement of his Abiogasco plant to handle 5,000 additional quintals a year; the expansion was completed in December 1963.

In 1964, Ricardo Remmele urgently requested the construction of a biologic plant on his farm *El Trevol*. The sanitation department would not allow the installation of his new *Astoria* sausage factory or the breeding and fattening of 2,000 swine on the same farm until he solved the problem of his factory and pigsty wastes which could pollute the downstream waters supplying towns. The construction was started in December 1964 and completed in March 1965. The plant would process the manure of 2,000 pigs of different ages and would be supplied with water from industry. The gas would be used for cooking and providing heat for the boilers.

In 1966, Rodolfo Castillo Love, owner and administrator of the farm *La Sierra*, requested the construction of a biologic plant for manufacturing biofertilizer exclusively. He was not interested in recovering the gas; (although it is always produced, the gas escapes into the atmosphere after passing through a pressure trap). The farm had a small hydroelectric plant and did not require further sources of energy. Its soil however needed to be replenished, as it had been planted with wheat for many years. The biofertilizer was ideal for the production of raw material, which was the manure of 50 partially stabled dairy cattle. This plant started operating in June 1966.

In 1966, the industrial company 'Biofert' (Alvarado, Losano, Penagos) was founded to produce biologic fertilizer for the domestic market. Estimated production was 50,000 quintals a year. Production started in 1969 and found a ready market in stores selling agro-chemical products. In 1971, demand exceeded production by more than 100,000 quintals, so compost was manufactured from pulp mixed with biofertilizer, both liquid and solid. The product was sold in thick, 50-pound polyethylene bags at US$1.50 each.

In 1969, Rosendo Gordillo G. requested the construction of a plant to manufacture biofertilizer only, in his *Rancho Alegre* farm in Rio Bravo, Escuintla. He was not interested in using the gas produced but wanted to process the manure from a stable of cows. The plant was installed at a distance from the farmhouse. The pasture soil was very sandy; in changing its texture with the use of biofertilizer, its water-retaining capacity was increased and fodder production improved. In 1969, economist Fernando Luna del Pinal requested the construction of a plant to produce biofertilizer on his farm *San Rafael Las Flores*, in the municipality of San Antonio, Suchitepequez. This plant was to be similar and operate under the same conditions as the *Rancho Alegre* plant; consequently, the 30 per cent additional investment for the installation to obtain biogas was unnecessary.

In 1977, the farmer and industrialist Milton Molina E. requested plans for a biologic plant for his *Las Acacias* agro-industrial installation in the municipality of La Gomera, Escuintla Province. He wanted to use the manure from a fattening stockyard of 8,000 head of cattle to obtain two basic products: biologic fertilizer and biogas. The fertilizer was for his cotton fields of sandy, clayish soil. The biogas was to be used as fuel in a large agro-industrial complex on the farm, which consisted of a cotton gin with a daily production of five hundred, 500-pound bales; a peanut oil-extracting factory; a cottonseed oil-extracting factory; and a factory using the cakes, husks and other industrial leftovers to manufacture 3,500 quintals of concentrate a day for stockyard fattening. Reference figures were the following:

(1) A total of 8,000 animals in rotation in open stockyards; 3,200 quintals of raw materials a day composed of equal amounts of manure and cotton haulm.
(2) With 8,000 head of cattle, the daily yield (on the basis of one cubic metre of biogas for each head of cattle) was 8,000 cubic metres of biogas.
(3) Combustion heat of the biogas was 23,800 BTU per cubic metre, a daily production of calorific energy of $200 \times 10^6$ BTU per day, equivalent to 1,680 US gallons of gasoline daily.
(4) Observation: Under the specific conditions, there would be a daily surplus of biogas for other uses, as follows:

| | |
|---|---|
| Daily production of energy | 200,000,000 BTU |
| Present consumption in plants | 114,660,000 BTU |
| Surplus | 85,340,000 BTU |

The amount of solid biofertilizer to be obtained would be approximately 90 per cent of the weight of the initial raw material; 10 per cent of the latter would be transformed into biogas. Consequently, the solid biofertilizer will be:

3,200 quintals $\frac{\text{(raw material)}}{\text{day}}$ x 0.90 = 2,800 quintals per day.

The amount of liquid biofertilizer is approximately 20 per cent of the weight of the initial raw material being processed. Consequently, the 'bioliquid' fertilizer will be:

3,200 quintals $\frac{\text{(raw material)}}{\text{day}}$ x 0.20 = 640 quintals equivalent to

$$\frac{640 \times 100 \times 1{,}000}{62.4 \times 35 \times 3.78} = 7{,}750 \text{ gallons per day.}$$

(5) Summary: From the 3,200 quintals of raw material processed (with 7 per cent initial moisture), the following would be obtained:

2,800 quintals of solid biofertilizer a day
7,750 gallons of liquid biofertilizer a day
8,000 cubic metres of biogas equivalent to 1,680 US gallons of gasoline.

(6) Note: In the biologic system, the amount of 40 pounds of pure nitrogen is

maintained for each head of cattle per year (in this particular case). Consequently, in a year, the 8,000 head of cattle would give 3,200 quintals of pure nitrogen, equivalent to 16,000 quintals of 20 per cent nitrogenized fertilizer. Furthermore, because of the handling and type of fermentation, the compost system results in a loss of 25 per cent organic matter. If this system is used, the daily loss would be equivalent to 800 quintals of organic matter.

On the basis of these figures, construction was started on a pilot plant to obtain detailed data on the variables of the raw material (weight, density, behaviour and temperature) as well as variables such as climate and moisture and mechanics of the soil. This pilot unit would be part of a battery of 18 bioreactors, all of the same capacity. The capacity of this bioreactor was 200 cubic metres. Its shape was hexagonal but this could be modified according to the behaviour of the raw material. The construction was completed on 28 December 1978, when it began operation, based on the estimated figures and taking into account the condition of the raw material, which was unsuitable at that time, being contaminated with beach sand. New loadings were made with the materials specified and with new materials, so as to start formal construction.

## EL PORVENIR PLANT IN SUCHITEPEQUEZ

### Description of the Use of its Products

**Solid Fertilizer:** Biologic fertilizer has been under production since the plant began to operate and only for the use of the *Margaritas* farm, located twenty kilometres to the north of *El Porvenir* farm where the biologic plant is located. The opinion of the owner was as follows: (1) the 'gut' wastes from cattle in a slaughterhouse located on the same land and possibly garbage from the market of the city of Mazatenango could be utilized; and (2) the truck that transports concentrates to cattle farms in the south will return with a load of biofertilizer and will also transport the pulp obtained from processing and, occasionally, manure from their farms. Three pounds of this fertilizer is used for each shrub and is placed on the upper part of sloped land and then covered with a vegetable mould.

During the first year of the experiments, the most spectacular case was the following: In 1961, there was an alarming outbreak of the coffee-leaf pest *Leucoptera coffeella;* the *Margaritas* farm was the centre of this outbreak. The coffee plants were just starting to be fertilized when damage to the leaves began; defoliation was progressive. The neighbouring farms decided to salvage all the coffee plants. In the end, only stumps shorter than one metre high remained. Inordinate amounts of chemical fertilizers were applied. The harvests in all the farms decreased substantially and, in some cases, there was a total loss. But a strange phenomenon was noted in *Margaritas:* as defoliation progressed, new leaves sprouted. The plants were not cut nor were huge quantities of chemicals used, and the harvest did not decrease. At the end of the year, the owner told me: 'My fertilizer plant was paid for during its first year because I did not lose my harvest as my neighbours did.' Another plague typical of the area is nematodes, especially *Pratylenchus* and *Heliocotylenchus* spp. The application of biofertilizers in liquid and solid forms seems to diminish their effect.

There was another experiment in 1962. In response to a suggestion, the owner, acting upon the idea of a well-known coffee grower from El Salvador (Ricardo Alvarez from Santa Ana), purchased a press to make compost pots. Biofertilizer pots were made; in the central empty cone of the pressed hexagon, more crumbled biofertilizer was added and the seed placed inside. The remarkable results obtained were:

(1) The 'damping-off' disease that makes the stalks rot was affecting the seed beds of the farms in the area, causing 40 to 60 per cent damage. In the seed bed planted directly on biofertilizer pots, there was not a single case of damping-off.

(2) The vigorous growth of the new plants made it possible to transplant the definitive small plant *eight months before* the usual time.

(3) Many people were interested in what was happening in the seed bed with the seedlings; they wanted to purchase the pots, together with the liquid or solid fertilizer. The owner refused all requests and ordered that the pot factory and its seedlings be moved inside the coffee plantation where no one could reach them.

**Liquid Fertilizer:** Liquid fertilizer was also applied to the coffee plants from the beginning. Some 4,000 US gallons a day were transported by tanker from the biogas plantation where two rows of coffee plants were sprayed with a polyethylene hose, several metres long, spreading an average of two US gallons around the base of each shrub.

The invigorating effect observed in these sprayed coffee plants was mainly in the growth of new leaves, flower verticils and new branches, which grow up to 50 and 60 per cent higher than normal. This effect was not measured in the coffee plantations but it was confirmed in later experiments.

**Biogas:** The use of biogas for the protection of *El Porvenir* plant is essential, since four of its motors are connected to the gas pipes. Nevertheless, each one of those motors can operate with gasoline whenever necessary because no changes have been made in their cylinder capacity nor in their carburation. The following problems were encountered:

(1) Gasoline must be used occasionally to start the motor when it is very cold, but it is immediately switched over to gas.

(2) Motors that have no internal lubrication and are lubricated by adding oil to the fuel (as is the case with two-stroke motors and very small motors in general) have these problems because the lubricant must be injected

when biogas is used.

(3) It was considered convenient to order the 'Wisconsin' motor, which is equipped with a carburetor for LPG, since biogas was unknown in the United States in 1960. But it was impossible to start the motor with all its gas equipment and it became necessary to remove the carburation system for LPG and make a simple, hand-operated device to solve the problem. The same was done in the other motors, which were smaller, such as the 'ONAN' electric generator and the 'REX' mud pump. The 'ILO' motor of the compressor already had an intake device for piped gas and its respective valve. The motors of a truck and an Allis Chalmers tractor were equipped with a device that was ordered for each, according to specifications, from the German company Hessenwerk Kassel.

It should be mentioned that the Wisconsin motor of the mill, which operated at least three hours a day for 12 years, did not suffer from carbonization or show valve damage; it was opened for a general checkup in 1973.

**Problems of Design, Construction, Operation and Use of Biogas Plants**

The study of a general analytic chart of biogas plants built in Guatemala, for the purpose of satisfying rural needs of small, medium and high economic level, gives an idea of the various problems involved:

**Design:** There is a specific design in each biogas plant built in Guatemala, covering the following features in particular:
  (1) the preference and wishes of the client who wants a simple fertilizer plant (28.5 per cent), a worthwhile installation (14.25 per cent) and a useful unit in his agricultural-livestock enterprise (43 per cent);
  (2) the location—whether it will be installed in the administrative centre of the farm (50 per cent) or on a distant site (50 per cent);
  (3) the use of the production—whether only biofertilizer (28.5 per cent) or biofertilizer and biogas (71.5 per cent);
  (4) the capacity: small, producing less than 45 cubic metres of gas per day (21.5 per cent); medium, between 45 and 100 cubic metres per day (64.25 per cent); and large, greater than 100 cubic metres of gas per day (14.25 per cent). In exceeding certain limits, movement of waste becomes complicated, in loading as well as unloading, and special transportation systems must be built;
  (5) the topography of the site, from very flat (28.5 per cent) to extremely uneven (71.5 per cent);
  (6) the climate, which will determine the protection to be provided against wind, cold and humidity. These factors affect yield. There have been constructions in temperate climates (50 per cent), in hot climates (43 per cent) and in cold climates (7 per cent).

In view of all these factors, the cylindrical design has been used in 21 per cent of the plants built in Guatemala. This design is the most suitable for economy of construction, interior cleanliness and hygiene. However, in the integration of a battery, this design causes problems because of the space remaining between the tangent cylinders, which causes humidity and loss of space. Furthermore, common walls cannot be used.

The rectangular design has been used in 7 per cent of the installations because it is less economical than the cylinder. Problems in cleaning interior angles have occurred only in one case. The problem of the city garbage—and consequently the sanitary aspect—would be solved with daily loading and unloading.

The system of circle segment has also been used in 7 per cent of the installations for the following reasons:

(1) a small overhead crane can easily introduce solids into each of the ten segments;
(2) the length of the pipe carrying the liquids pumped is substantially shortened;
(3) the extraction of the finished products is economical and easy, by unloading the finished biofertilizer into a single receptacle, situated in the interior of the segments, where its liquid is poured off by gravity into an interior pit; similarly, the biogas is collected from each of the segments in a single pipe leading to the gaśometers;
(4) there is a saving in construction of walls; one common wall can serve for every two segmented tanks;
(5) the hexagonal shape is the one most used (64 per cent) because of its special features, which make it similar to the circle: many common walls are used, the interior angles are open and suited to a cellular battery formation with different grouping forms.

**Construction:** In regard to construction, only one plant has been built with baked mud bricks, with cement plaster in the interior and small frames on the walls. Problems of leakage, longer construction time and shorter durability have occurred in only 7 per cent of the installations. Only one plant, *Chipo*, was built with steel. The owner preferred this because the plant could be transported from one place to another.

On the other hand, reinforced concrete was used in 86 per cent of existing plants. It was preferred because of the facility and speed of construction, resulting from the use of modular forms, and the durability and flexibility in certain details of the design. The materials for concrete are easily obtained throughout the country, cement and iron being the only foreign materials obtained. Labour was readily available, either on the same farms or in nearby towns.

**Operation:** Operation is perhaps the most important aspect and depends directly on the system designed, the type of raw material used and the volume of production. The system designed for a single load of homogenous material, with volumes not greater than 45 cubic metres of load per day, can be handled by one or two unskilled people. This is the case in 50 per cent of the plants built in Guatemala; these may be called simple operations.

An installation with a capacity greater than 45 cubic metres per day requires an auxiliary mechanical or hydraulic transport-type of loading and unloading system. Its operation requires a plant foreman and two labourers. Twenty-one and one-half per cent of the installations built are of this type. Semi-industrial or industrial plants with a capacity greater than 100 cubic metres per day (whether for a single loading or daily loading and unloading), require more sophisticated mechanical transport equipment, pumps and instruments. Skilled workers are needed including electricians and laboratory technicians. The construction of biogas plants in our medium should be simple to operate. There have been no major problems when the personnel handling them have been trained and follow-up has been maintained.

**Use of the Products:** The use made of the products—whether the raw materials or finished products, both biofertilizer and biogas—involve the following problems:
  (1) The raw materials, composed of substances having little polymerization, are easy to hydrolyze and the most recommended biologic system for their degradation is high dilution.
  (2) Highly polymerized materials, on the other hand, decompose more slowly; the system most recommended for their anaerobic degradation would consequently be of very low dilution.
  (3) Many raw materials necessarily require aerobic pre-fermentation before being subjected to methanizing fermentation.

The same consideration should be taken into account with regard to the use made of the finished products. For example, when the biofertilizer is going to be used on the same farm, the appearance of the finished product is unimportant. If, on the other hand, it is going to be marketed, it should be finely disintegrated, ground and packaged. Similarly, biogas should be filtered if it is to be used in motors of internal combustion; its carbon dioxide content should be no greater than 35 per cent for a good yield and operation of the motors. Biogas to be used in burners, stoves, dryers and other forms of direct combustion require no treatment.

Such considerations have arisen in every case where interest has been expressed in constructing biogas plants and, in each case, the problem has been solved as best as possible, taking into consideration the other factors.

## CONCLUSIONS

### (1) Diversity of Needs

**Sanitary**: Sanitary needs have arisen only in those cases where the respective authorities or a pressing need has made the use of a treatment system necessary, which has occurred in 14 per cent of cases. Nevertheless, this feature is one of the most important to be considered. In China, it has been necessary to install millions of small family and community biogas plants. In Guatemala, one such plant is in operation promoted by Centro Mesoamericano de Tecnologia Aplicada (CEMAT), and is proving popular among the Indian population of San Pedro La Laguna, Atitlan; an effort is being made to improve yield and facilitate unloading.

**Fertilizer**: The use of fertilizers of organic origin has become more common recently, especially in the United States, where demand is greatly increasing. In Europe, their use has been and continues to be of vital importance. In our medium, the Indian considers it to be the vitality of the land; in mountain zones, fertilizers are always used when planting corn.

One hundred per cent of the demand for biogas plants in Guatemala has been for the purpose of using biofertilizer, either by the owner or for sale to third parties. For 71.5 per cent, it is used as the main product of the plant and the biogas is a byproduct; 36 per cent use it as the only product, with the biogas escaping into the atmosphere.

**Energy**: Only since petroleum has become so expensive has the demand for gas become important; in our experience, there was only a 28.5 per cent demand at the beginning. At present, most of the requests are for biogas, although the biofertilizer has not lost importance.

### (2) Different Areas

**Urban Areas:** In urban areas, the most immediately available supply of energy is the cities' sewage waters, which at present are not used at all and cause pollution, disease and foul odours. In the new developments skirting the cities, sewage waters and rain water could already be separated. The energy that could be produced through generated biogas would cover peak electric requirements and pump water from the deep wells of the municipality. In the near future, the installation of a metered supply of energy to homes, industries and services, through compressed biogas in large pressure tanks, could be considered.

**Rural Areas:** These areas have the largest demand (79 per cent) because it is easy to obtain good-quality wastes. With further development, biogas would bring the following positive results to rural areas: better hygiene, prevention of diseases and promotion of community development with the participation of co-operatives. It would also enrich the soil with excellent organic matter. It should be noted that 'fertilizers obtained as byproducts' are more useful to the farmer than fuel because it is still possible to obtain and transport firewood, in spite of increasing difficulties.

**The Domestic Sector:** It uses energy mostly for caloric purposes (cooking), water heating, direct mechanical purposes (water pumps) and direct lighting. Conversion to electricity is also being considered.

**The Productive Sector:** It uses energy mainly for irrigation, stationary combustion motors, electric generators, water pumps and mobile agricultural machinery. In this sector, the use of biogas is also complemented by domestic uses and must be shared by milk-producing establishments, stockyards for the fattening of cattle, pig-breeding, poultry-breeding and vegetable-growing, since this is the best alternative from the sanitary point of view, the least expensive and the most profitable. Such is the case of the Astoria Sausage Factory in Guatemala.

**The Agro-industrial Sector:** It uses biogas to produce heat for industrial processing, to generate electricity and to dry agricultural products. In Guatemala, 'Finteca' attempted to introduce biogas to dry coffee throughout Central America; some demand already exists. In this sector, the success of biogas depends mainly on the availability of wastes suitable for processing or on the location of the industry in relation to other centres that could supply raw materials.

## (3) Diversity of Raw Materials

In the 21 years that have elapsed since 1958, the diversity of raw materials has been an important aspect of the utilization of new available materials obtained from agro-industrial processes:
   (1) In *El Porvenir* plant, utilization of coffee pulp was requested. The same request was made for the *San Alberto, Colima* and *La Esperanza* plants. The results were completely successful (1958-1959).
   (2) The municipality of Guatemala undertook the treatment of market garbage in 1959, with pre-fermentation of the materials.
   (3) In the *Chipo* plant, experiments were made for the first time in the utilization of sugar cane bagasse, with very successful results (1960).
   (4) In the *El Porvenir* plant, experiments were made to utilize *citronella* pulp and lemon 'grass', but no positive results were obtained.
   (5) In 1981, the Bionomic Resources Company of New York commissioned us to perform research on the utilization of sawdust from trees growing in the northern part of the state: beech, red oak and sugar maple. This

research had originally been requested by the New York Botanical Gardens and the Cari Arboretum. In February 1982, experiments with that material were begun and one of the three different treatments applied showed great promise. It produced 250 litres of biogas per kilogram of sawdust from beech ash, which equals the yields of cereal straw.

## RECOMMENDATIONS

(1) Interested institutions, researchers and owners of sources of organic wastes should carry out systematic research on native raw materials.

(2) Simple and economic digestors should be designed to solve the tedious and permanent control of the pH in the environment and suitable technology should be used for manual unloading after a period of several months.

(3) Promotion and development agencies should grant soft credits for the construction of digestors in rural areas.

(4) This new source of energy and humus should be considered a complement of the dairy herds, the stockyard and agro-industrial processes.

### References

ICAITI (1956) Bibliographic Report. Reference 54/7/01, 6 July 1956

# III Production of Solar Collectors in Mexico

**Fernando O. Monasterio**
Universidad Autonoma Metropolitana Xochimilco, Mexico

THE INDUSTRIAL manufacturing and commercialization of solar water heaters has been developed on a small scale in some Mexican cities, especially in Guadalajara, Jal., where such were initiated 35 years ago and most recently in Cuernavaca, Mor., Mexicali, B.C. and Mexico, D.F.

### Uses of Solar Heaters in Mexico

In Mexico solar water heaters are built to be used for domestic or industrial use and for swimming pools. The domestic use requires water at 40°C, making it necessary to use the effect of greenhouses in the solar heater, with which temperatures of 60°C and higher could be obtained. The traditional arrangement consists of a flat collector formed by a metallic plate and tubes where the water passes, all of these located in a box, thermally insulated and covered with glass, and in a tank for storing hot water, also thermally insulated.

In the case of the swimming pools, it is necessary to heat the water to temperatures between 24°C and 28°C. These temperatures could be reached, in high insolation zones, with a solar heater without using the effect of the greenhouse, which has caused some manufacturers to simplify the design and to reduce the cost by suppressing the glass. The solar heater is reduced to copper or aluminium plates painted in black or oxidized and copper tubes attached to the plates where the swimming pool water circulates by force of a pump.

### Development of Solar Water Heaters in Mexico

In 1942, in Guadalajara, the Solar Heaters company was created and is still in production. Up to 1977 it had completed over 2,000 installations, especially in Guadalajara, Jal., but also in La Pieded, Queretaro, Culiacan, Saltillo, Monterrey, Cuernavaca and Mexico, D.F.

A reduced domestic market was initially developed for solar water heaters for clients of the upper-middle class; but the applications diversified for heating swimming pools and for some industrial water-heating installations, for public baths as well as for agricultural uses.

The use of solar heaters has increased especially in Guadalajara, where there are presently various manufacturers. The interest in solar heaters increased in recent years and new manufacturers have appeared, particularly in Cuernavaca, Mexicali and Mexico D.F. The market has widened, principally the one for swimming pool heaters. Recent designs are supported by technical literature published in the United States as well as in some research done in Mexico, and are the most efficient and economic.

With regard to solar water heaters for domestic use, the present tendency is to combine them with gas heaters, which has allowed a saving of about 60 per cent in gas consumption.

## Industrial and Commercial Aspects

Information related to technical characteristics and prices of two manufacturers in Guadalajara, one in Cuernavaca and two in the Federal District, was obtained as it was considered a significant sample.

Solar Heaters (Industrias Orozco Carricarte, S.A.), in Guadalajara manufactures solar water heaters with capacities of 200 litres to 1,000 litres. Prices range from 10,000 Pesos for the 200-litre capacity up to 39,000 Pesos for the 1,000-litre capacity. The company has 40 employees, including those in charge of installation and periodical check-ups.

Proveedora de Calentadores Solares, also in Guadalajara, manufactures heaters with capacities from 100 litres at a price of 5,500 Pesos up to 1,000 litres at 28,000 Pesos. The materials used for manufacturing are sheets of galvanized steel for the collector's box and the thermo-collector tank of steel or copper plates, glass cover and polystyrene insulation. The company has constructed an economical prototype of 110 litres at a price of 6,000 Pesos, which includes collector and thermo tank.

Solarmex, S.A. de C.V., in Mexico D.F, manufactures the solar water heater for domestic use (Scotimagas). The company uses a copper tube and aluminium slats or copper slats painted in black. The collector is covered by a glass 6 mm thick and insulated with fibreglass. The thermo tank is manufactured by Calorex (manufacturers of gas heaters). With a capturer of 0.86 x 0.56 metres and a deposit of 115 litres, the Scotimagas has a price of 7,250 Pesos; with two capturers and one tank of 240 litres, the price 14,144 Pesos.

## Life of Heaters

In Mexico there have been solar water heaters in use for more than 15 years. Manufacturers consider a solar water heater to have a useful life of more than 20 years.*

---

* The Mexican heaters were installed by Calentadores Solares and Enrique Ramoneda; information granted by Eng. J. Manuel Orozco, General Manager of Calentadores Solares; Eng. Octavio Garcia, Manager of Modulo Solar; Eng. Enrique Ramoneda, Manager of ITESA and by the representative of SOLARMEX, SA.

## Amortization of the Investment

Various manufacturers estimate that the investment in a water heater could be amortized in 4 to 5 years, based on the saving of fuel at the present prices of domestic gas. In an economic analysis performed by the Engineering Institute of UNAM (Fernandez Zayas *et al,* 1977), the following conclusions were drawn:

(1) Whenever comparing solar alternatives with the traditional ones of combustion, the costs of fuel are outlined as the most important. The fuel represents 80 per cent of the total cost of the alternative of the circulating gas heater and 88 per cent of that corresponding to the deposit gas heater, in a period of 15 years.

(2) The use of solar energy is much more economical than the gas equipment, regardless of the efficiency of the latter. For instance, the present value of the use in 15 years of a conventional solar heater, although initially it has higher costs, is only one-third of the value of a circulating gas heater with high efficiency.

(3) The largest initial investment corresponding to a solar heater is recovered in one to two years from the saving in fuel, although the financing cost could be high.

## Response of the Local industry

All the companies that manufacture solar water heaters are industries with a small manufacturing level using nationally manufactured materials.

Up to this date no important industrial organization has entered this field, although (according to the Engineering Institute of UNAM), Vidrio Plano de Mexico, manufacturers of glass of Monterrey, and Houbard and Bourlon Enterprise, dedicated to making electrical installations for air conditioning, have intentions of branching out into this field.

## Institutional Response

The manufacturers of solar water heaters have not had any stimulation or help from the authorities. On the other hand, the government has financed costly projects importing foreign technology (Proyecto Tonatiuh), which has not provided any benefits to the country.

## Future Perspectives

It is surprising that a small market for solar heaters has been developed in Mexico without any official stimulation, considering how cheap liquified gas is from petroleum (domestic gas). Regarding the immediate future, according to one of the manufacturers the market of solar heaters for swimming pools after a fast initial growth will continue to grow slowly. In order to enlarge the market of solar water heaters for domestic use so that the lower-middle class, which is a wide potential market, can benefit, it is necessary to develop an inexpensive heater that could be purchased in a hardware store (as are gas heaters), and that any plumber could install without making major modifications to the existing installation.

The increase in the price of domestic gas, which seems to be unavoidable, will contribute to enlarging the market, especially if authorities adopt policies that will help finance the purchase of solar water heaters or allow such purchases to be tax deductible. A policy of this type could allow, in the short term, important savings in consumption of domestic gas to accumulate and in this way will contribute to the conservation of non-renewable resources.

### References

Fernandez Zayas, J.L., Garibay, J.J. and Gonzalez, A. (1977) *Water Heaters for Domestic Use.* Engineering Institute, UNAM.

# IV National Alcohol Programme in Brazil

Secretariat of Industrial Technology,
Ministry of Industries and Commerce

THE NATIONAL ALCOHOL PROGRAMME was created by Decree No. 76.593 of 14 November 1975 in order to satisfy the requirements of the internal and external markets and the motor vehicle fuel policy. The decree provides that the production of alcohol made from sugar cane, mandioc or any other input shall be promoted by expanding the supply of raw materials; special emphasis will be given to increasing agricultural production, modernizing and enlarging existing distilleries and to installing new production units attached to the independent factories, as well as to storage units. An increase of three billion litres in alcohol production in Brazil to obtain a mixture of 20 per cent alcohol with the gasoline used in the country in 1980 (the goal established for the first stage of the National Alcohol Programme) should be obtained during the 1978 − 1979 harvest. This goal is due in part to the international sugar crisis which has favoured the use of idle distilleries and the processing of the enormous sugar cane surplus into alcohol.

Present technological developments guarantee the technical and economic feasibility of alcohol as an alternative source of energy to petroleum by-products, either for mixtures or for direct use in converted or manufactured motors.

For the above reasons, the government undertook a comprehensive evaluation of the programme that resulted in the proposal for new goals and the subsequent expansion and intensification of the activities of PROALCOOL. The goal of producing 10.7 billion litres of alcohol in 1985 was to yield the following possible utilization:
(1) 6.1 billion litres of hydrated alcohol for 1,700,000 vehicles of which 1,225,000 have factory motors and the remaining 475,000 have converted motors;
(2) 3.1 billion litres of anhydrous alcohol to be added to gasoline (20 per cent);
(3) 1.5 billion litres of alcohol for the alcohol-chemical industry.

Nevertheless, this goal could be increased up to 20.5 billion litres of alcohol, depending on the behaviour of oil prices and technological developments in the application of alcohol. This effort should already be connected with the programmes anticipated for a second stage through the following increase in the alcohol production capacity (from 9.8 billion litres of alcohol): (a) 4.8 billion litres of alcohol for partial substitution of diesel oil, amounting to 14 per cent of the total and (b) 5.0 billion litres of alcohol for partial substitution of fuel oil or for a more substantial substitution of gasoline.

To attain these goals, it is necessary to implement, research and develop projects that will contribute to improving efficiency in the areas of agricultural inputs, industrialization, treatment of wastes and the utilization of alcohol.

It is important to point out that in the industrial process, the technologies that will be evaluated and adapted will make it possible to reduce the cost of invested capital. Research on the uses of alcohol and its effect on motor technology and the performance of converted motors as well as on the industrial utilization of alcohol should be intensified. For the implementation of a programme of this scope, approximately 3 per cent of the resources invested by PROALCOOL are allocated to research.

It is necessary to make some considerations resulting from a socio-economic evaluation carried out by PROALCOOL.

**Economy of Foreign Currencies:** By mixing 1.5 billion litres of alcohol with the gasoline used in 1978, PROALCOOL made possible a small economy of foreign currencies, favoured by adjustments in the structure of the refining process and the exportation of gasoline surpluses. Nevertheless, assuming a production of 20.5 billion litres of alcohol in 1985, the economy of foreign currencies would amount to US$ 24.5 billion for the 1979-1985 period. Besides this, it would also produce earnings of US$ 3.8 billion for exportation of gasoline surpluses. The investment required to implement this programme is approximately US$ 8 billion.

**Reduction of Regional Differences in Income:** Although participation by the north-eastern region in national alcohol production increased from 17 per cent to 29 per cent, the largest investments (approximately 65 per cent) were allocated to the central-southern region. It is necessary to re-examine the financial incentives to favour the most needy regions of the country.

**Reduction of Individual Differences in Income:** An agro-industrial programme of the magnitude of PROALCOOL would bring new population groups into the market economy by giving purchasing power to deprived groups of the population. However, to attain an effective reduction of income differences it will be necessary to adopt other correlated economic policy measures, since the mere increase of employment within a salary structure tending to the concentration of income could, on the contrary, strengthen it further.

**Increase of Internal Income:** The 218 distilleries proposed by the National Alcohol Council (CNAL) provide investments, at constant February 1979 prices, of some 43.6 billion cruzeiros. Considering that almost all of these investments will be made internally, in cruzeiros, through the purchase of equipment and other industrial and agricultural tools (most of which are produced in Brazil) and through the construction of industrial facilities and agricultural subsidies involving entirely national resources (construction materials, land and labour), the contribution of PROALCOOL to the increase of internal income due to the multiplying effect of those investments in the economy becomes evident.

**Expansion of the Capital Goods Industry:** The industrial investments of PROALCOOL are aimed at modernizing, establishing and enlarging distilleries. Approximately 65 per cent of these investments are assigned to the purchase of equipment that, at constant February 1979 prices, amounts to almost 23.5 billion cruzeiros. These investments are causing considerable pressure on the equipment demand, which is being met by the Brazilian industrial park. It has been noted, however, that attaining this objective has been detrimental to another objective of the programme, that of reducing regional income differences, since the resources channeled for the purchase of equipment are concentrated in the state of Sao Paulo, where the two main manufacturers are located.

**Institutional Mechanisms**

The activities involved in the operation of PROALCOOL provide for joint efforts by government agencies and national private enterprise, under the co-ordination of the National Alcohol Council (CNAL). With the exception of the CNAL, participating public agencies existed before the programme and it was necessary only to increase duties and resources to meet new requirements. Production, distribution and commercialization continue to be handled by the private sector under the new financial incentives established by PROALCOOL, especially in the agricultural and industrial sectors. Technological solutions for the production of raw materials (particularly of sugar cane), equipment and of the end-product already existed when the programme was launched. It is therefore clear that the strategy adopted has been based on the granting of financial incentives, on the utilization of available organizational and technological resources and on decentralization of the decision-making process, while CNAL has assumed the general co-ordination of activities. In view of the multiple agencies and institutions participating in the development of activities and of the characteristics of the institutional framework of PROALCOOL, the sphere of action of the programme has been divided into four phases.

**Phase 1. Raw Material Production:** The constant drop in sugar prices on the international market during PROALCOOL's existence made available a full supply of sugar cane, the main raw material used for alcohol production. On the contrary, the programme helped to overcome the sugar crisis by processing into alcohol (through a government decision) a considerable amount of raw material used to produce sugar. In view of this elasticity in the use of sugar cane resulting from the international situation, the activities of the programme have been channeled to promoting production and commercialization of alcohol. The government policy, as regards the production of raw material, has been practically limited to granting financial incentives to ensure sugar production through the National Rural Credit System and to research aimed at increasing the saccharose content in sugar cane and the development of alternative crops (mandioc, babassu palm, sugar sorghum, cellulose) and products that could contribute to the substitution of the other oil by-products, under the co-ordination of the Secretariat for Industrial Technology (STI).

**Phase 2. Alcohol Production:** In the phase of alcohol production, the co-ordination, regulation, control and financing activities of the government are carried out by various agencies within their own spheres of competence. The main activities undertaken are the following:

(1) The National Alcohol Council (CNAL) defines criteria, analyzes and organizes proposals for modernizing, enlarging or establishing alcohol distilleries.

(2) The Sugar and Alcohol Institute (IAA) determines alcohol prices, defines production types and volumes, establishes technical specifications of the different types of alcohol, provides technical and administrative support to CNAL, gives technical assistance to the production process and to the financial agents for PROALCOOL, supervises alcohol production and acts in the field of research through PLANALSUZAR.

(3) The Secretariat for Industrial Technology (STI) co-ordinates research carried out by different public and private institutions on the development of processes and industrial equipment and the treatment of wastes.

(4) The Special Secretariat for the Environment (SEMA) determines guidelines and undertakes control of water pollution caused by industrial wastes in internal waters.

(5) The National Monetary Council (CMN) establishes financing conditions and the annual amounts of resources to finance projects and their warranties.

(6) The Financial Agencies (AF) study the economic and financial feasibility of projects, carry out credit operations and supervise the

execution of projects.

(7) The Central Bank (BACEN) regulates the industrial and agricultural credit operations, generates financial resources and refinances agro-industrial resources.

**Phase 3. Distribution of Production:** Through Decree No. 82.476 of 23 October 1978, the activities of the distribution and commercialization stages of alcohol production for fuels were transferred to the enterprises distributing petroleum derivatives, under the guidance, regulation and supervision of CNP, together with IAA, on those aspects that fall within the competence of this agency. This new mechanism was regulated through Resolution No. 18/78 of the CNP of 11 November 1978.

**Phase 4. Alcohol Commercialization/Utilization:** Given the technological limitations existing at the time PROALCOOL was created, its first stage calls for the utilization of alcohol only as a mixture for gasoline. Since then, new applications have been submitted to research and technological developments by the government through the Programme for Alternative Renewable Energy Sources of Vegetable Origin and by the motor industries.

The main areas that are being developed for the utilization of alcohol are conversion of gasoline motors to use hydrated alcohol, diesel oil substitution and alcohol chemistry. At the institutional level, the government action in the field of alcohol utilization (as well as in the other technological aspects) has been developed under the supervision and co-ordination of the STI.

**Legislation and Acts Regulating Alcohol**

The first legislative and regulatory acts related to the alcohol industry in Brazil were issued in 1931. It should be pointed out that the IAA, through Resolution No. 146 of 25 September 1974 created a commission to study a new version of Law No. 5.998 of 10 November 1943. The results obtained by the commission have not yet been analyzed by CNAL.

**The National Alcohol Council and National Executive Alcohol Committee**

In view of the numerous public and private institutions participating in PROALCOOL, co-ordination is essential, especially in the decision-making process. At the time PROALCOOL was created, the National Alcohol Commission was established as an institution for collective deliberation in order to avoid the duplication of efforts and deviations from the objectives and to ensure the fulfilment of all the basic functions by the various institutions.

As a result of the evaluation of PROALCOOL undertaken by the government in 1978, it was decided to change the National Alcohol

Commission into the National Alcohol Council (CNAL) by Decree No. 83.700 of 5 July 1979, so that it could formulate policies and determine the guidelines of the programme. This decree also created the National Executive Alcohol Commission (CENAL).

# V  Charcoal Siderurgy in Argentina

*M.A. Trossero*
OLADE Consultant

PIG-IRON PRODUCTION constitutes one of the most important industrial sectors of this country's life. In the past, siderurgical industry carried out its first steps using charcoal as a fuel and reducing agent. As time passed, this industry began to evolve according to the tendencies and situations in which the world was living in the technico-economic context; thus charcoal siderurgy transferred to coal, perhaps for the following reasons:

(1) rapid decrease in forest reserves;
(2) growing demand for steel, which generated great siderurgical complexes with blast furnaces of ever-increasing capacity;
(3) increased availability of coking coals;
(4) improvement of coking techniques and exploitation of coal mines.

This situation evolved favourably from the 1930s to the 1970s, when a certain weakness of the existing structure began to appear because of the heterogeneous distribution of coking coal deposits, and the growing cost of fuels.

The panorama generically described has led Latin America to face the situation from another perspective, as may be seen from the diverse meetings and conventions held by ILAFA which came to the following conclusions (ILAFA-CARBON, 1976; Vieira, 1977; IBS, 1976 a, b; Thibau, 1972):

(1) regional scarcity of coking coals deposits;
(2) lack of the necessary substructure for utilizing the most important mineral resources of the region (Colombia's case);
(3) growing international cost of coke;
(4) developing technologies for incorporating new reductors will not be applicable before the end of the century.

Within this rather sombre panorama, vegetable resources as siderurgical reductors seem to be most favoured, at least in the Latin American context, for the following reasons:

(1) vegetable resources are renewable;
(2) the region counts on great natural mountains, suitable lands, enough sun

and water for reforesting to establish a charcoal siderurgy;
(3) improvement of charcoal blast furnaces technology;
(4) processes of pig-iron production of charcoal, competitive with coke.

This situation enables one to foresee that siderurgical charcoal industries tend to consolidate themselves in their markets, and increased production should be expected by enlarging the existing ones and/or by new projects (ILAFA-CARBON, 1976; Vieira, 1977).

## ARGENTINA

CHARCOAL SIDERURGY in Argentina is represented by Zapla Blast Furnaces which appeared around 1945, when the first pig-iron was produced using national raw materials, and then complemented by the construction of a steel plant, hot roll and the Forest Centre for charcoal production.

The evolution of pig-iron production in Argentina may be seen in Table 7.3 being presently 240,000 tons per year, with the inauguration of Blast Furnace No. 5 with an estimated consumption of charcoal of approximately 300,000 to 320,000 tons per year. Presently, pig-iron production is increased by Tamet, SA, with a charcoal blast furnace whose production of pig-iron is approximately 50,000 tons per year.

Charcoal production in Argentina, according to statistics, averaged 450,000 tons per year during a five-year period, from which around 250,000 to 300,000 tons per year is used for the home market. If the mentioned siderurgical consumption capacity is considered to be 350,000 tons per year, would there be a charcoal scarcity nowadays? The fact is that siderurgical plants are not performing at full capacity. If they were to do so, it is estimated that there is a potential production capacity of charcoal that may respond with a certain agility to the growing demand.

## BRAZIL

CHARCOAL SIDERURGY in Brazil has a long history, beginning with the arrival of the Royal Portuguese Family in 1808, who participated in several local undertakings without much significance, until Belgo Mineira Company settled down in 1921, making a contribution of approximately half of the country's production of steel around 1930-1940 (Vieira, 1977).

Presently, Brazil relies on twelve charcoal-integrated industries and more than one hundred non-integrated ones (Table 7.4). The energy situation the world is now facing leads to a review of the uses of forest resources, which, in the short and medium term, will have an important role in the search for alternative energy sources.

Table 7.3  Evolution of Pig-iron Production in some countries of Latin America ($\times 10^3$ tons)

| Description | 1960 | 1965 | 1970 | 1975 | 1977 | 1985 |
|---|---|---|---|---|---|---|
| ARGENTINA | | | | | | |
| Blast furnaces—charcoal | 23.7 | 73.6 | 98.0 | 76.9 | 113.6 | 240.0 |
| Blast furnaces—coke | 120.7 | 589.6 | 712.3 | 961.0 | 986.0 | 9600.0 |
| Pig-iron production | | | | | 1385.2 | |
| BRAZIL | | | | | | |
| Blast furnaces—charcoal | 966.2 | 1043.4 | 1743.0 | 3477.7 | 3664.0 | 6400.0 |
| Blast furnaces—coke | 792.8 | 1407.5 | 2327.3 | 3384.7 | 5537.2 | 18920.0 |
| Pig-iron production | | | | | 9764.6 | |
| LATIN AMERICA | | | | | | |
| Blast furnaces—charcoal | 1026.3 | 1079.5 | 1841.9 | 3554.6 | 3777.6 | 7000.0 |
| Blast furnaces—coke | 2024.9 | 3447.2 | 5479.0 | 7414.6 | 10417.2 | 52810.0 |
| Pig-iron production | | | | 12814.2 | 16867.1 | 60420.0 |

Sources: IBS—Comissao do materias primas— Datos ostadisticos, 1977
ILAFA - 18- Memoria Técnica, Congreso Latinoamericano de Siderurgia
ILAFA - Anuario Estadistico - 1967 and 1971
Siderurgia Latinoamericana—April 1978, No. 216

**Table 7.4 Evaluation of Pig-iron Production by Charcoal in Brazil**

| Brazil | 1967 | 1968 | 1969 | 1970 | 1971 | 1972 | 1973 | 1974 | 1975 | 1976 |
|---|---|---|---|---|---|---|---|---|---|---|
| Total pig-iron —charcoal | 1,196 | 1,296 | 1,655 | 1,896 | 2,177 | 2,310 | 2,461 | 2,535 | 3,346 | 3,777 |
| Total pig-iron | 3,078 | 3,228 | 3,766 | 4,235 | 4,743 | 5,287 | 5,533 | 5,858 | 7,016 | 8,174 |
| Share (percentage) | 38.9 | 40.1 | 44.0 | 44.8 | 45.9 | 43.7 | 44.5 | 43.3 | 47.7 | 46.2 |

Analysis of planned siderurgical projects of different Latin American governments suggests that soon new countries of the region will turn to charcoal for siderurgical production, since this means the assumption of independent plans due to the ever-increasing cost of imported coal and coke.

Latin American governments would do well to consider national siderurgical alternatives in view of the politico-economic situation that is producing the present energy crisis. The situation will be favoured by a greater effort of national and private enterprises to develop with more speed the necessary technology for steel production by means of charcoal. In the meantime, development implementation and the investigation concerning charcoal manufacture, which has old and rudimentary techniques, will be necessary.

**Argentina:** Charcoal siderurgical enterprises existing in Argentina estimate that they can maintain present pig-iron production capacities until 1985; in the case of Altos Hornos Zapla, a new fine rolling check gear is being completed. The Tamet project expects to double its production by 1981 (Tables 7.5 and 7.6).

Table 7.5    Charcoal-Siderurgical Enterprises

| Country | Enterprise | No. of Blast Furnaces | Annual Pig-iron Production (tons) |
|---|---|---|---|
| Argentina | Altos Hornos Zapla | 5 | 240,000 |
|  | Altos Hornos Güemes | 1 | 12,000 |
| Australia | Wundowie | 2 | 70,000 |
| Brazil | Diversas | 136 | 3,777,400 |
| Great Britain | Backbarow | 1 | — |
| Russia | Nodishdruck | 1 | 65,000 |
|  | Verknyeyasalda | 1 | — |
| Malaysia | Malayewata Steel Bnd. | 2 | 130,000 |
| Thailand | Tho Siam Ironaud Stell Co. Ltd. SISCO | 3 | 60,000 |

Table 7.6   Projects of Charcoal - Siderurgical Enterprises in Latin America

| Country | Enterprise | Annual Pig-iron Productions (tons) |
|---|---|---|
| Argentina | Tamet* | 100,000 |
| Brazil | Varias | 100,000 |
| Bolivia |  | 100,000 |
| Honduras |  | 100,000 |
| Paraguay | Acopar* | 100,000 |
| Uruguay | Valentinos | 100,000 |

*In construction

**Brazil:** It is a country having a large charcoal siderurgy due to its large output of 3 million tons per year, and to the technological development reached in the last years. Moreover, growth programmes predict 5.5 million tons per year for 1985. The programme will be complemented by studies and implementations of large areas for forest exploitation to produce the desired charcoal levels (Table 7.4).

**Latin America:** At present there are several projects, some of them near completion; soon they will be integrated with the existing ones. Several projects worth mentioning include (Table 7.6):

(1) Paraguay: with ACEPAR SA, an integrated plant with a 100,000 tons per year capacity is under construction;
(2) Uruguay: in the region of Mina Valentines, an integrated plant of 100,000 tons per year of nonflat rollings is projected;
(3) Bolivia: Honduras, Peru and Chile are analyzing similar projects.

The described panorama indicates a tendency towards a moderate growth of this industry (see Table 7.3), which in two or three years will be even more favourable due to the critical international panorama.

## Technologies of Carbonization

The firewood distillation and carbonization industry counted on diverse systems and methods throughout its history. It may be said that each coal region had its own system; at the time, producers introduced particular innovations that did not much improve the processes themselves. In spite of the prevailing situation, methods used were defined by the aims pursued: industries used retorts and their corresponding recovery equipment for firewood distillation; and stack systems or masonry furnaces for carbonization.

The diversity of systems produced uncertainty about the most convenient selection. This situation prevails at the present time, bringing up several discussions from adherents to one system or other for contrary reasons rather than technical justifications. A classification of carbonization furnaces is worked out in Figure 7.2.

Regarding constructive characteristics the reader is referred to the bibliography at the end of this section (Vieira; Kollmann; Mariller; Klar; Earl; Trossero; CVD).

**Carbonization Furnace**: This hemispherical furnace (Media Naranja) is constructed with bricks only (10,000), settled on mud. It has two opposite doors for operation of firewood loading and discharge of charcoal. The most common furnaces have the following dimensions:

| Cycle Duration (days) | Diameter (m) | Theoretical Volume (m$^3$) | Firewood Volume (m$^3$) | Charcoal Production (kg) |
|---|---|---|---|---|
| 15 | 7 | 90 | 60 | 7,000 |
| 10 | 6 | 60 | 40 | 5,000 |

Hemispherical furnaces are discontinued production types, and are operated in groups of 7 to 10, conducted by a burner, an assistant and four carriers. The energy for transforming firewood into charcoal comes from the load itself. The process of carbonization is made from top to bottom. This type of furnace is widespread in Argentina.

**Brazilian Furnace**: This furnace has a cylindrically shaped surface, with arched roof, constructed with brick settled on mud and reinforced with a metal band. It has two opposite doors, sometimes metallic. The main dimensions are generally: diameter, 5 metres; height, 3.6 metres with a theoretical volume of 55 cubic metres; load, 41 cubic metres of firewood; and produces approximately 5,000 kg of charcoal per cycle, which lasts from 8 to 10 days.

**Raw Materials**

Generally, it may be said that any vegetable matter can be used for producing charcoal, but this is produced in places where there is demand by the siderurgical industry, which requires a suitable quality compatible with the performance of the blast furnace.

This leads to the utilization of the thickest portions of the tree (trunk and some branches) for charcoal manufacturing. The situation described generally occurs not because the manufacturing activity offers a good yield, but rather as a means of covering the cost of land-clearing prior to agriculture and cattle

CASE STUDIES: CHARCOAL SIDERURGY IN ARGENTINA 127

Figure 7.2 Classification of Carbonization Furnaces

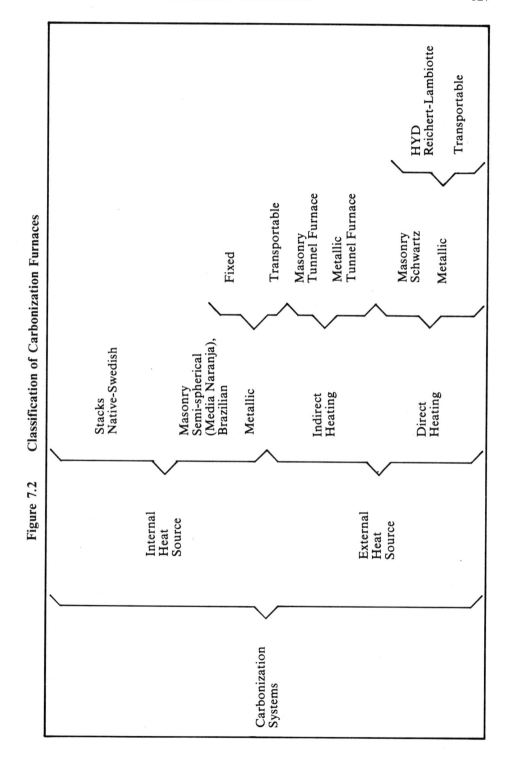

production; the type of mountain that is generally degraded is constituted by species of trees that in many cases have a low commercial value in the traditional market. This is the case of Monte Chaqueno in the north of Argentina, where hard and semi-hard woods are found.

For this reason, coal production in Argentina is secondary, since it is just the outcome of agriculture and cattle frontier expansion. Reforestation for charcoal production is nonexistent, except for the Centro Forestal de Altos Hornos Zapla, which has 10,000 hectares of eucalyptus and production plants capable of approximately 1,500 tons of charcoal per month by means of Brazilian-type furnaces.

Another example to be considered is the enterprise Salta Forestal S.A., which with rational and integral management of the original mountain (taking out old and dead wood, leaving renewables and young trees for the mountain reconstitution) keeps the original features of the ecosystem. Its semi-spherical furnaces have a production capacity of 1,500 tons of charcoal per month.

Until a short time ago, the panorama in Brazil was similar to that of Argentina. Beginning with the official decision of facing a programme to reach 6 million tons per year of pig-iron using charcoal by 1985, legislation was worked out, forcing siderurgical enterprises to generate their own forest resources so that by 1985 they should produce at least 50 per cent of the charcoal utilized in each siderurgical plant.

Reforestation is carried out with diverse varieties of eucalyptus (*grandis, rostrata, saligna, tiriticornis*) which have a seven-year cycle (the cycle is expected to be reduced to five years), and produce from 15 to 30 tons per hectare of vegetable matter each year, i.e., an equivalent of 4-10 tons per hectare of charcoal a year.

The scheme, brought about by siderurgical enterprises in conjunction with the Brazilian government, becomes easily feasible and may be worth mentioning as an example: a siderurgical enterprise producing pig-iron of 100,000 tons per year requires two years for its construction and can be supplied by charcoal coming from 20,000 to 30,000 hectares of eucalyptus. The inconvenience may arise from the implementation time difference of the siderurgical plant with regard to forest plantation; nevertheless, it may be avoided by the utilization of original mountains of the region for producing charcoal, and as long as the earthwork is performed, it may be reforested with new, more suitable species for the pursued aim.

The cost for implementing such a scheme is estimated to be about US $500-600 per hectare, including a soil cost of approximately 10 per cent of that cost, which would prove to be an implementation cost of approximately US $10,000.

**Manufacturing Costs of Charcoal**

The most important factors contributing to the charcoal conformation cost, produced by means of a hemi-spherical furnace (Media Naranja) and metallic for comparative effects, are generically described as follows:

|  | N.M.H. | Metr. H. |
|---|---|---|
| Cutting, round-up and transport to furnace | 10.00 | 10.00 |
| Charcoal manufacture | 15.00 | 7.00 |
| Amortization of equipment | 3.00 | 17.00 |
| Freight and movements | 15.00 | 8.00 |
| Losses by fire | 3.50 | 3.50 |
| Miscellaneous expenses | 3.50 | 5.00 |
| Total | $ 50.00 | $ 50.50 |

It was observed that resulting costs of the utilization of hemi-spherical (Media Naranja) and/or metallic furnaces are similar, with the advantage that the latter system results in better quality and uniformity, and better working conditions for workers.

The cost of reforesting the cleared land surface with eucalyptus is estimated at present to be US$500 to 600 per hectare; there is approximately a 50 per cent discount for land and felling costs. Thus, an equivalent of US $25 per ton should be added to the manufacturing cost of charcoal (US $35 per ton of charcoal) in order to reforest the cleared land surface and, in this way, restore the biomass that will generate at the same time 10 to 30 tons per hectare of charcoal per year.

**Buying Price of Charcoal**

The buying price of charcoal at present, by Altos Hornos Zapla, is $100 per ton from the factory. TAMET SA is paying $70 per ton from the factory. This is to say that presently the estimated cost from the factory will be US $50-70. These prices are similar to those existing in Brazil.

**Comparative Costs of Reductors**

The yield of charcoal price in pig-iron production is estimated to be 50 per cent to 60 per cent, and from 35 per cent to 40 per cent in the steel produced. A comparative calculation is shown as follows, in order to point out the possibilities of siderurgy run by charcoal with respect to the traditional coke:

$$\frac{\text{Cost of pig-iron run by coke}}{\text{Cost of pig-iron run by charcoal}} =$$

$$\frac{\text{US\$140} \times t^{-1} \text{ coke} \times 0.75 \text{ t coke} \times t^{-1} \text{ pig-iron}}{\text{US\$70} \times t^{-1} \text{ charcoal} \times 1.0 \text{ t charcoal} \times t^{-1} \text{ pig-iron}} =$$

$$\text{price relations of reductors} = \frac{\text{Cost of coke}}{\text{Cost of charcoal}} =$$

$$\frac{\text{US\$140} \times t^{-1} \text{ coke}}{\text{US\$70} \times 6^{-1} \text{ charcoal}} = 2.0$$

## CONCLUSIONS

FROM THE preceding analysis, one can see that siderurgy run by charcoal has at present an outstanding yield in pig-iron production in Latin America.

|  | 1976 | 1985 |
|---|---|---|
| Siderurgy run by charcoal | 27% | 9% |
| Total production of pig-iron in Latin America | 59% | 72% |

It is observed that forecasts indicate a decrease of charcoal in the region, but $3.9 \times 10^6$ tons of pig-iron produced in 1976, and $6.7 \times 10^6$ tons expected to be produced in 1985, are amounts that may cover all siderurgical requirements of a country like Argentina.

Enlarging native and unexploited zones with abundant forest resources and low commercial value species may provide a suitable yield in the charcoal production for supplying the siderurgical industry. A plant of 100,000 tons of pig-iron per year requires a biomass surface of approximately 25,000 hectares, which at the same time generates some 10,000 new jobs, a favourable factor to be considered in the future of the Latin American region.

The scheme described has been studied extensively with regard to its technical and economic feasibility; it is competitive with the coke alternative, and allows small plants to be supplied with a renewable resource of national origin. As a consequence and in conclusion, the extensive experiments carried out by Argentina and Brazil must be imitated and perfected. It is hoped that biomass will play an important role not only in the future generation of energy but also in supplying an extremely valuable renewable siderurgical reductor.

## References

Borges, Colombarolli (1978) *Carvao Vegetal, Opcao Energetico para a Siderurgia dos Paises Tropicais.* Congreso ILAFA, Altos Hornos '78, Buenos Aires.

CVD (1953) *Los Hornos que utiliza CVD en 'Los Tigres'.* Dept. de Agronomiá, Buenos Aires.

Earl, D. E (1977) *Informe sobre Carbón Vegetal.* Congreso ILAFA 18, Memoria Tecnica, November 1977.

IBS (1976a) *Siderurgia: Debato major emprego dos Carvao Vegetal.* IBS Revista, No. 14.

IBS (1976b) *Carvao em Debate.* IBS Revista, September.

ILAFA—CARBON (1976) *Usos del Carbón en Siderurgia, Abastecimiento y Tecnologia.* Mexico, July.

Klar, M. (1977) *Technologie de la Distillation du Bois.* Congreso ILAFA 18, Memoria Tecnica, November 1977.

Kollmann, F. (1977) *Technologien des Holzes und der Holzwerkstoffe.* Congreso ILAFA 18, Memoria Tecnica, November 1977.

Mariller, C. (1977) *La Carbonisation du Bois, Lignites et Tourbes.* Congreso ILAFA 18, Memoria Tecnica, November 1977.

Thibau, C.E. (1972) *Abastecimiento de Carvao Vegetal a Siderurgia Brasileira.* Sid. Latino-americana No. 150, October.

Trossero, M.A. (1978) *Analisis Comprativo de Hornos de Carbón Vegetal.* Congreso ILAFA, Altos Hornos '78, Buenos Aires.

Vieira, O.P. (1977) *Tecnología do Carvao para a Siderurgia na America Latina.* Congreso ILAFA 18, Memoria Tecnica, November.

# VI Las Gaviotas: An adequate Technological Centre that functions*

*Jose Miguel Velloso*
Ceres, FAO

IN COLOMBIA, the region of Los Llanos is a large, almost uninhabited savannah, extending towards Venezuela and Brazil. Las Gaviotas, the Centre of Integrated Technology, is found in this region, installed in an old abandoned military camp. It is not easy to get there from Colombia's interior. One must cross the highlands, with peaks of over 4,000 metres, and then travel across many kilometres of almost desert savannah. One can get there by jeep as far as Villavicencio; this takes about three and a half hours and then, a plane trip of an hour and a half brings one to Las Gaviotas. The easiest way to arrive there, however, is by plane directly from Bogota, which takes two and a half hours.

From the air, one can see all of Los Llanos, crossed by numerous water streams—a solitary region with an exuberant vegetation. The population is small and most settlements are found close to large patches of burnt land—the result of local efforts to fertilize the earth; some settlements can also be found near the streams. Large areas of burning grass indicate places where devastating, as well as ephemeral, cultivation has been practised. Other ochre-bare spots indicate burnt places that were plotted and then abandoned after being productive for a while; these spots will probably never regenerate.

Nowadays, Las Gaviotas has a hospital, school, workshops, mess halls, living quarters, vegetable gardens and cattle. The Centre was created about ten years ago, thanks to Paolo Lugari, a Colombian sociologist, who had the idea of

---

* This text originally appeared in *Ceres,* FAO's journal on Agriculture and Development (Volume 11, Number 3, May/June 1978) under the title *Las Gaviotas: The Centre of Integrated Technology where everything has been proved on site.*

researching the possibilities for developing this large, non-productive zone. From the beginning, Lugari had the collaboration of the Colombian government and other institutions, among them the University of Los Andes and Professor Jorge Zapp (Dean of the Mechanical Engineering Faculty of the same university), a mathematician and physicist who soon became the technological brain behind the project.

Today, Las Gaviotas is a study centre dedicated to one of the most interesting technologies in the world. Three fundamental principles have ruled its functioning from the beginning: to go ahead, step by step, without 'stretching your arm further than your sleeve'; not to take anything for granted without experimenting first; and to be limited to sectors where evidently nothing concrete has been done, accepting and using the available tools or machines that have demonstrated their efficiency. The ruling philosophy of Las Gaviotas is: 'There is no reason to go back to the Stone Age whenever trying to create an appropriate technology. But all the knowledge that humans have acquired to manufacture machines and tools, as a result of a very complicated technology, yet of easy utilization and maintenance, should be used.'

Notwithstanding the poverty of its soil, the possibilities of developing Los Llanos are enormous. Some of them could be carried out immediately. Others require certain environmental changes that, although difficult, are not impossible. One of these, for instance, is the raising of cattle, for which it was necessary to find a certain type of beast that could survive on the poor grass of the savannah. Two types of animals were found: the Zebu and a type of African sheep similar to the goat, but one that would not impoverish the land. Around the Centre, one can now find large herds that give meat and fresh milk. Another long-term possibility would be reforestation of the zone, which could be done with a variety of pine: the *Caribea*. If this reforestation could be done, these pines could provide Colombia with the necessary cellulose not only for the country's paper necessities but also for exportation. This possibility is being studied at Las Gaviotas in the corresponding institutions.

Regarding the most immediate possibilities of exploitation, there are two types of palm trees (*selje* and *moriche*) which grow well near the rivers and from which a yellow, pleasant-tasting food oil could be extracted. Such oil could be industrialized and sold in sufficient quantities to cover a large demand. But the greatest problem is to repopulate the area and provide the people with simple and cheap equipment to cultivate the land.

Las Gaviotas gives an answer to this through a concourse of its researchers, who have ventured to do the extraordinary: listen to what common sense advises. In other words, the researchers themselves have decided to settle in the savannah and to face the problems that such a settlement would have. Laboratory studies, physical theories, mechanical knowledge—these are the elements they use. Their solutions are the results of immediate and repeated experiments, under environmental conditions and real physiological stress, after months and years of effort to provide the necessary efficiency and practical application to reach their goals. All materials to be used would be local or, at least, cheap and easy to obtain.

## CASE STUDIES: LAS GAVIOTAS TECHNOLOGY CENTRE

The history of Las Gaviotas is one of intentions, tests and rectifications. The researchers' own construction of the buildings for different services and for housing were to be from local materials. First it was necessary to solve one of the most difficult problems of construction, that of locating an efficient and cheap roofing that would last. Los Llanos has tropical weather and, for this reason, the roofing has to allow a good interior temperature, while at the same time be resistant to heavy rains. The conventional systems were experimented with and the present buildings have straw roofs with a large chamber of air and without an internal ceiling, allowing ventilation. But this does not seem to be the ideal solution since, although this is a cheap type of roofing, it has to be renewed at least every six years. It seems that the project researchers may select clay shingles for future roofing.

The researchers of Las Gaviotas have correctly believed it to be senseless to look towards areas which have already been sufficiently researched and where conclusions and concrete fulfilment have been obtained. One of their tasks is to acquire those tools or machines of adequate technology to serve their purposes, or to construct others for which descriptions and designs could be obtained. Surprisingly, they have found that, in a majority of cases, these machines do not work, function badly or frequently get damaged.

One machine which Las Gaviotas has begun to produce industrially is the 'yucca grater'. In the tropical zone of South America, as in Africa, the yucca grows and is widely consumed. Aside from the nutritive value of this product, which is limited, the starch obtained has great commercial value in the plastics industry. The Centre studied a yucca cutter of African origin that was moved by pedals (like those of a bicycle) and they discovered that what needed to be done was not to cut the yucca but to grate it. The location and type of knife did not permit this kind of operation, nor did it prove efficient. Therefore, necessary changes were made and knives were converted into saws, the location angle was changed and finally the present grater was obtained. It has been functioning for some years now and is able to produce 500 to 1,000 kg of yucca pulp per day. This grater needs only two people to operate it, one to pedal and another to insert the yucca.

Water is the first necessity for any human settlement. It has to be pumped from wells or streams to take it to reservoirs, troughs or canals. For this reason, one of the first concerns of the Centre has been to obtain machines that will pump efficiently and cheaply the necessary water for households and the community.

Las Gaviotas is full of windmills of all origins; the Centre wanted to acquire and experiment with them to prove their efficiency, handling and durability. None of them could be adapted to the operation that, in theory, had to be performed. Aside from the high cost, their complicated mechanism broke down easily, their fin blades could not handle the wind properly and, consequently, their efficiency was low. Then the Centre decided to manufacture their own windmill which was impressive in its simplicity, cost, firmness and efficiency. With cloth fin blades, a tail that guides it to take the slightest wind and a proved resistance to violent winds, this windmill can pump from 7 to 20 cubic metres per day with a pumping altitude of 2 to 25 metres. Its resistance is proved by years of functioning and its cost of US$400 is lower than the cost of any other windmill of similar characteristics presently on the market.

With the same purpose of pumping water, but this time for domestic use, the Centre created an umbrella ram and a manual injection pump of surprising simplicity and efficiency. Various types of umbrella rams can be found on the market. But all of them have a defect, aside from rapid deterioration: they must be regulated very carefully according to the water fall. The ram produced at Las Gaviotas is self-regulated and functions simply, thanks to a physical law related to the water pressure. Its efficiency is 1 to 3 cubic metres daily, with an altitude of 4 to 25 metres using falls of 1.5 to 3 metres. At a cost of US$25, it is six times less than the lowest priced ram currently found on the market. To give an idea of the care given to the testing of tools and machines at Las Gaviotas, it will be enough to mention that the prototype of the umbrella ram was discarded because it was discovered that children at play could put their fingers inside it, and cut themselves. In order to avoid this risk, the umbrella ram was modified.

The injection pump can be easily handled by a young boy of 12-15 years old, thanks to the design of the shaft that moves it which requires less strength than was originally needed. Such a pump—useful for filling the watering places of cattle, for irrigating small plots and for other domestic utilizations—has a pumping altitude of 2 to 8 metres with a flow of 0.5 to 2 hectolitres per second. Its cost (US $15) is less than one-third of the cheapest similar model found on the market.

The Centre then turned its attention to the water-pumping problem and the production of energy for collective and industrial utilization. Since Los Llanos is a practically flat region, the water streams found there do not have high enough falls to be able to establish conventional hydroelectric plants. Therefore the problem of generating hydroelectric energy in the region is compounded: on one hand, the water has to be recaptured to create a high enough fall to move the turbine and, on the other hand, a certain type of turbine has to be found that can function efficiently with water falls of modest levels.

Firstly, Las Gaviotas created a plastic bag with perforations, filled with fast-setting cement. The bags, a little heavy, are piled up in the deep part of any water current until a dam is formed. The cement sets on contact with the water and the bags remain stacked due to their own weight as well as the cement. In this way, dams are formed that will produce high enough falls to operate a horizontal turbine constructed in Las Gaviotas to solve the second problem. This dam feeds a conventional transformer that is available on the market.

A dam constructed like this has another important advantage: it is both solid and flexible (something similar to the shield of an armadillo) and avoids breaking due to increased water pressure or land movement, as in the case of the rigid dams. There is a hydroelectric plant of this type functioning at the Centre which could power all the machinery of its workshops.

To provide enough electrical energy to the average household (with electrical illumination such as refrigerator, television, radio and iron) the Centre invented an axial microturbine, a model of which is installed and has been functioning for some months now. This microturbine, no larger than a one-kilogram tomato can, requires a flow of 60 to 70 litres of water per second and functions with a fall of one metre and one-half to two-and-a-half metres, with a potential of 700 to 1,000 watts. It is connected to a conventional transformer and could produce sufficient electricity for a normal household or for six to ten rural houses with limited electrical appliances. Its cost (US $150) is twenty times less than a turbine of similar nature on the market.

The hot water installed in houses, hospitals, and so on is obtained through a solar heater manufactured also at Las Gaviotas, with a cost that represents one-third of any other similar device on the market. It supplies 100 litres of water per day at 65° C.

The Centre has also successfully experimented with a solar heater that uses spent fluorescent lighting tubes as raw material. A team of young boys gathers these tubes. Their cost is practically nil and they are delivered to the Centre's headquarters in Bogota, where they are conveniently prepared. This type of solar heater has an enormous advantage in being built *in situ* and with the help of instructions sent from the Centre can be easily assembled.

With regard to agricultural cultivation and the raising of poultry and other animals, the Centre has adopted a technology, with which experimentation has been done in Brazil, consisting of the establishment of greenhouses where pots with one cubic metre of soil to which small quantities of elements that are lacking have been added. In this way there is not the waste which would be involved in trying to distribute those elements over large land areas, and vegetable plants and products that are difficult to cultivate in tropical areas—for instance, the onion— are obtained.

In addition to that received from the Colombian government, the Centre also receives support from the United Nations Development Programme (UNDP) and, more recently, from the Government of the Netherlands. The Centre favours the idea of an international congress of technology to exchange ideas. All performances of Las Gaviotas are available to people throughout the world. The Centre is planning to start industrial production of the machines and tools mentioned so far; the cement bags for dams are already being produced. All machines and tools are adapted to similar conditions as those at Los Llanos.

The researchers of Las Gaviotas insist on the following: 'We will only accept tools and machines that have been proved on land and not laboratory illusions that, in reality, cannot be transformed into useful things. Because what functions in theory does not always function in reality.'

# VII The Present State of Geothermy in El Salvador

*Gustavo Cuellar*

Co-ordinator of the Geothermy Project, OLADE

EL SALVADOR, the smallest republic of Central America, has limited hydraulic resources and, because of its geological make-up, limited possibilities of being able to include oil byproducts within the scope of its natural resources. However, this country has succeeded in alleviating its national energy problem and is on the way to becoming completely independent in electricity generation by using the steam coming from its great subterranean heat reserves.

In 1966, the first investigations in this field were systematically initiated by the Comision Ejecutiva Hidroelectrica (CEL) of Rio Lempa with the technical assistance of UNDP. In the first phase of investigations, geoscientific studies and shallow exploratory perforations were carried out in areas which showed signs of superficial heat. Subsequently, five deep exploratory wells were constructed. The results of the explorations in these wells determined the pattern of development in the second phase of the programme.

The second phase of investigations was concentrated in the geothermic region of Ahuachapan. It began in 1969 and ended in 1971. This phase included detailed geoscientific studies, the perforation of five deep wells for production and a feasibility study to construct a generator of 30 MW. In 1972, CEL continued its geothermic studies in Ahuachapan, establishing the techno-economic feasibility to install a central geothermic plant of 90 - 100 MW. In 1975, the first geothermic unit of 30 MW began to function; in June 1976, a second unit of equal capacity was installed; a third unit began operation at the end of the first term of 1980.

During the development and implementation of the geothermic region of Ahuachapan for industrial operations, the following investigations were being realized:

(a) large-scale re-injections of residual geothermic water with the aim of optimizing exploitation of reservoirs and minimizing the impact on the surroundings;

(b) investigations in the geothermic region of Berlin, reaching the point of doing a feasibility study for the installation of a fourth geothermic unit of 55 MW;

(c) investigations in the geothermic regions of Chinameca and San Vicente, reaching the pre-feasibility phase;

(d) first phase of studies to determine the geothermic potential of high and low thermodynamic functioning in the whole country, with a view to its industrial benefits;

(e) training seminars for specialists in geothermics at Central American level.

## REGIONAL GEOLOGICAL CONDITIONS

ONE-THIRD of the Salvadorean territory is of the Pleistocene-Holocene age; the remaining two-thirds are of the Pliocene-Miocene (Tertiary) age. All the formations corresponding to those periods are of a volcanic nature and cover the underlying sedimentary rocks of Cretaceous-Jurassic age; but, due to the limited refining area, these formations do not play a significant role within the geological framework. To the north of the country, there are formations of intrusive rocks of a granite nature belonging to the Miocene age. The younger volcanic rocks (Pleistocene-Holocene) offer better temporal conditions for geothermic development in the case of El Salvador.

The more characteristic structural element is made up of a central graben which runs through the entire country in a west-northwest/east-southeast direction. This graben also continues into Guatemala to the northwest and Nicaragua to the southeast. The coastal and interior chains constitute the marginal block of the graben. During the Pleistocene age, a type of volcanic activity was developed related to the geological faults, thus giving rise to this structure, principally on the southern border where volcanic strata rose, partially covering the internal part of the graben.

The geothermic regions identified with high temperatures (200-325°C) are found along this structure and can be seen during the process of exploitation or development. Parallel to this, in the north of the national territory associated with volcanoes degraded due to erosion and which are completely inactive, there exist geothermic zones of average temperature (approximately 150-200°C).

## THE GEOTHERMIC REGION OF AHUACHAPAN

IN THE NORTHEAST region of the volcanic group Laguna Verde and on the southern margin of the graben is the geothermic area of Ahuachapan. The west-northwest/east-southeast and northeast/southwest systems characterize the whole central area. However, recent tectonic activity (to which volcanoes are tied) reveal signs of thermals and post-volcanic phenomena in the study area, which result in this area being controlled by a younger system, north-northwest/south-southeast.

**The Quaternary Eruptive Centres:** Laguna Verde and Laguna Las Ninfas constitute the heat source of the region and the principal recharge of the reservoir; the base of the region is made up of a type of agglomerated water-proof rock belonging to the Tertiary volcanism. The rocks of the reservoir are made up of a series of high-potency laval strata, basals of andesites of Ahuachapan—underneath which is a succession of lava and calcareous tufa. The geothermic system of Ahuachapan is embraced within an area of approximately 200 square kilometres; however, the operations area of the region is approximately 2.5 square kilometres.

The total number of perforated wells is 30, equivalent to 29,000 metres of perforations, of which 16 are productive with a minimal distance of 150-160 metres between them. Of the remaining wells, four are for re-injection and ten for explorations—but some of the latter are of limited productive value. The exploratory wells are spread out over an area of 8 square kilometres. The producing wells form an anchor pipeline of $13\frac{3}{8}$ inches, precipitated from 0-100 metres deep; a pipeline of production of $9\frac{5}{8}$ inches precipitated from 0-500 metres deep (inferior limit of the roof of the reservoir); a pipeline with a groove of $7\frac{5}{8}$ inches or a free opening of $8\frac{1}{2}$ inches to correspond to massive agglomerates that receive the re-injection fluids. The depth of the perforated wells in Ahuachapan vary between 591 and 1,524 metres; the mechanical finish is determined by the stratigraphic characteristics, the distribution of permeability and the temperature.

After initiating the exploitation phase in the geothermic regions of Ahuachapan, by means of an equilibrated system of extraction/re-injection, a careful programme of measurements and observations has been developed that allowed the following conclusions to be drawn:

(1) During the period of development of the region (1968-1975), 21,390 kilotons of geothermic fluid were produced. Simultaneously with the changes in extraction, some important changes were observed in the pressure of the reservoir, which confirmed that the reservoir has a natural recharger on the outskirts of the extraction zone.

(2) During the period June 1975 to December 1978, the operation of generation units implied a total production of 18,500 kilotons of geothermic fluid per year, reaching an accumulative total of 75,100 kilotons during the period 1968 to the end of 1978.

(3) As a result of the decrease in pressure observed during the exploitation period of the region, a steam zone was developed in the high structures of the reservoir, thus generating an increase in the proportion of steam produced by the wells (50 per cent), in comparison with the original steam production (15-20 per cent).

(4) The re-injection of residual water showed a definite benefit for the maintenance of the reservoir pressure, especially when levels of the order of 50 per cent of the extracted mass are re-injected. The combination of natural recharge and re-injection permitted the recuperation of some levels of pressure in the reservoir.

Additionally, the noxious effects on the surroundings have been minimized.

The following simplified table shows the units of average pressure in operation and of double pressure in the process of installation. The two existing units operate with steam of average pressure. Its nominal capacity is 30 MW for a consumption of 230 tons per hour.

The third unit has been predicted to operate with steam of both average and low pressure. The steam of average pressure will be obtained by the early separation of the mixture in the head of the wells. The water separated at this point will be directed towards the vaporizers located near the plant, which operate at 1.65 kg/cm² abs. The characteristics of the turbines and conditions of operation are as follows:

| | First and Second Units | | | |
|---|---|---|---|---|
| Condition of Operation | Charge | Flow of Steam (T/h) | Pressure put on Turbine (kg/cm²)abs | Empty Condensator (kg/cm²)abs |
| Nominal | 30 MW | 230.0 | 6.0 | 0.085 |
| Maximum | 34.5 MW | 261.5 | 7.0 | 0.0965 |
| | Third Unit | | | |
| Condition of Operation | Low Steam Pressure (T/h) | Average Steam Pressure (T/h) | Throttle Valve Pressure M.P. (kg/cm²) | Throttle Valve Pressure B.P. (kg/cm²)abs |
| A: 35 MW | 145 | 171.02 | 5.59 | 1.53 |
| B: 35 MW | 170 | 155.24 | 5.13 | 1.57 |
| C: 40 MW | 170 | 186.98 | 6.12 | 1.73 |

The 60 MW corresponding to the two 30MW units installed in the geothermic plant of Ahuachapan correspond to 14.3 per cent of the amount installed in the country. Its operation has been completely satisfactory, with an availability factor of 95 per cent. Since June 1975, when the first unit began to function, until August 1979, 1,414,585 MW/H had been generated. This is equivalent to 114 million Bunker C gallons in the proportion of 12.4 KW/H gallons according to the characteristics of the local thermal plants.

The percentage contribution of geothermic energy to that generated at a national level reached 32.3 per cent annually. The maintenance of the plant has been programmed for every two years, with a duration of approximately one month for the maintenance of the wells and superficial installation, which are carried out alternately without interrupting the operation of the central plant.

## GEOTHERMIC PROJECT — ORIENTAL ZONE

THE GEOTHERMIC PROJECT of the Oriental zone of the country begun in 1976, covers 2,500 square kilometres and integrates under one umbrella the areas of Berlin, San Vicente and Chinameca. As a result of preliminary studies, geothermic anomalies were identified underground and these are presently being evaluated to determine their dimension and potential.

### Geothermic Region of Berlin

This geothermic region is located on the northeastern slope of the volcanic group by the same name. It is directly associated with the recent volcanic activity of cones, craters and laval flows developed in the interior of the boiler of the primitive Berlin volcano, whose relative age has been estimated as belonging to the Pleistocene period. During the final phase of volcanic activity, this boiler was modified by tectonic effects, being divided according to parallel gravitational faults in a north-northwest/south-southeast direction, which gave rise to another secondary structure with graben-like characteristics. The geothermic region has developed precisely to correspond to the structures described—boiler and sand—which together cover an area of approximately 100 square kilometres.

However, the geothermic anomaly defined in superficial investigations is less than that indicated by the geological structures involved, but with a very narrow relationship. The geological model of the region has been preliminarily constructed on the basis of lithological profiles obtained in the perforation of deep exploratory wells. According to samples collected, the formation of the reservoir is made up of fractured andesitic lavae.

The well Tronador 2, the first perforated during the present programme, is 1,902 metres deep. Its productive zone was found at 1,799 metres; the maximum temperature measured was 310°C at contour 900. Samples of flow taken indicated a production of 100 kg/sec of the water-steam mixture with approximately 40 per cent steam.

The geothermic region of Berlin presents favourable conditions for its economic exploitation. Its geostructural conditions as well as its thermodynamic characteristics indicate the presence of a reservoir of high thermodynamic functioning underground—principally of a liquid type. Data collected suggest a minimum serviceable potential of 1,110 MW.

A pre-feasibility, techno-economic analysis was done for the installation of a central plant of 55 MW—the first phase of operation being 1984-1985. This includes a thermodynamic analysis to determine the necessities of steam for the generation and the system of installations that offer better advantages for the Berlin region that are of the turbine type with a double flow of condensation.

Based on the experience acquired in the exploration and exploitation of geothermic regions, the necessary investment has been estimated in order to pursue the exploitation phase at the Berlin Camp. The considered expenses include investigations, perforations of deep productive and non-productive wells and the acquisition and installation of equipment. Expressed in millions of dollars:

| | |
|---|---:|
| Turbogenerator | 16.0 |
| Pipelines | 1.0 |
| Electrical equipment and auxiliaries | 1.3 |
| Perforations and studies | 14.0 |
| Civil Work | 4.2 |
| Land | 1.0 |
| Residue | 0.9 |
| Engineering and Administration | 4.1 |
| Contingencies | 3.8 |
| TOTAL | 46.3 |

**Geothermic Region of San Vicente**

San Vicente is developing on the northern slope of the volcanic apparatus, with a twinned cone-like structure in whose craters a subtle fumarolic activity takes place. This fairly recent apparatus is seated upon a boiler of Tertiary age. The final effusive phase of the volcano consisted in flows of a type of andesitic-dacitic lavae. The superficial geothermic activity located on defective zones is genetically connected with the heat source of the volcano. The geological-structural scheme defined on the basis of geophysical data indicate the underground presence of an eventual structure of the Horst type which coincides with one of the two geothermic anomalies detected. The deep exploratory well of San Vicente 1, in the process of perforation, has intercepted (at a depth of 1,000 metres) a zone with temperatures higher than 200°C and with good conditions of permeability, thus confirming the existence of a new reservoir that could be exploited in the immediate future. At present, detailed investigations and deep perforations of exploration are being continued, with the aim of evaluating the zone's characteristics and potential.

**Geothermic Region of Chinameca**

Chinameca is located on the northern skirts of the volcanic apparatus. The activity of the group is in two eruptive phases: the first corresponds to the origin of the primitive Chinameca volcano in the Pleistocene period; the second is of a more recent period—it came about as a consequence of the tectonic collapse of the primitive volcanic structure, which was divided by a secondary graben on a north-northwest/south-southeast bearing. In the centre of this graben, a new volcanic cone emerged whose crater was called Laguna El Pacayal. The geological scheme of the region is favourable for the formation of geothermic reservoirs. The lithological profiles of the perforated wells indicate the existence of autosello formed by the hydrothermic alteration of pre-existing volcanic rocks. The reservoirs intercepted by the deep exploratory perforations are integrated by partially silicified andesitic lavae.

According to geochemical indicators, temperatures of 195°C can be surpassed by means of new locations near the source of the heat. However, the intercepted fluids in the middle run can be made profitable using non-conventional systems of energy conversion. Investigations of the zone continue.

## INVESTIGATIONS IN NEW AREAS

CONSIDERING that up to now the geothermic experiences have been very satisfactory and that they are of great importance for the country, the ability to develop the geothermic potential in an accelerated manner (especially for the high absolute growth in demand which obliges the hydroelectric plants to operate in the peak of the system) makes necessary the implementation of geothermic units as a base. It thus requires a unit of 55 MW every two years, for which all efforts ought to be concentrated to obtain information that defines the real potential of geothermic resources in El Salvador.

Considering the specific characteristics of each region, as well as the lack of a universal method to resolve problems related to the distinct phases of investigating geothermic exploration and automatically allowing the identification and evaluation of a region, the application of an adequate methodology has a more delicate aspect in the selection and combination of techniques that tend to reach particular objectives at every phase of investigation. In fact, the wide variety of possible local conditions and the geo-scientific criteria of specialists can exert substantial changes in the sequence and characteristics of the techniques of exploration anticipated.

However, with a philosophy of exploration, supported in local and foreign experiences, OLADE (as a contribution of exceptional form to the Latin American Geothermic Community) has succeeded in initiating the implementation of the most adequate methodology for the development of geothermic projects in distinct stages and according to the reality of the Latin American countries. This methodology is being incorporated widely into the exploratory programmes, especially in the phases of recognition and pre-feasibility. For El Salvador, in the short term, additional investigations in new zones ought to be initiated, which would permit a definition of the national geothermic potential. A minimum of 500 MW is presently estimated in the geothermic regions of high temperature already studied. Probably a similar potential can be assumed in the low temperature areas, which can be of benefit by using binary cycles.

The evaluation of new areas will be initiated gradually. The first stage will be the lateral and central parts of the departments of San Miguel and La Union. The objectives to be followed include: evaluating in a preliminary form the geothermic possibilities of the region studied, selecting areas of interest and elaborating a preliminary geothermic scheme and the systematic programme of exploration for each phase.

## HUMAN RESOURCES

ONE OF THE PRINCIPAL factors that can limit the benefit to be derived from the exploration and exploitation of geothermic potential of the country at regional level is the scarcity of specialized manpower. Some countries have tried to solve this problem temporarily by contracting consultation firms. However, experience has shown that because of the very nature of the programme to be developed:
(1) it is difficult to find a truly competent firm that can substitute the contribution of qualified personnel;
(2) the time required is much longer and the costs are considerably higher;
(3) the training of local professionals has been generally underestimated.

International organizations, such as UNDP and OLADE, have developed activities in the hope of collaborating with countries interested in the use of geothermic resources, promoting meetings and seminars with the objective of capacitation, because of the lack of means available to satisfy the demand for training at Latin American level. Since November 1978, the Republic of El Salvador, with the co-operation of the United Nations as part of the Central American energy programme, has held a series of seminars to train professionals coming from Central American countries, with the hope of South American and Caribbean countries eventually participating. The programme of seminars consisted of:
(1) 1st Seminar: Planning and Methodology in the Development and Exploitation of Geothermic Resources—General Orientation (November–December 1978)
(2) 2nd Seminar: Techniques of Geological, Geochemical and Geophysical Exploration (June 1979)
(3) 3rd Seminar: Perforation of Geothermic Wells and Measures (February 1980)
(4) 4th Seminar: Engineering of Reservoirs and Evaluation of Resources (June 1980)

# VIII Drying of Coffee with Renewable Sources of Energy

*Alberto C. Alas*
Superintendent of Investigation, Hydroelectric Commission
for Rio Lempa, El Salvador

COFFEE is an extremely important agricultural product for El Salvador. The coffee dryers that are most commonly used in this country consume approximately one gallon of diesel for every hundred-weight of coffee; 12 per cent of this weight is the moisture that results from the process. In terms of actual price, this quantity is equal to a net expenditure of about 1.20 Salvadorean Colones for every hundred-weight of coffee. The cost of transportation and the quantity lost in drying the coffee are not included in this price. In other words, for every million hundred-weight of coffee that must be dried, it is necessary to spend 1.2 million Colones on fuel, assuming that it would all be dried by the same heating methods. These amounts cannot be ignored by a small country such as El Salvador.

## Limitations

Let us refer to the coffee that is prepared by using the water process: water carries the coffee bean through the successive stages of peeling, fermentation and washing (Coste, 1960). Let us also assume that large quantities of ripe coffee are brought daily to the plant during the coffee harvest and that there are large amounts of pulp and skin as a waste product of this process. Finally, let us assume that the coffee harvest coincides with the dry season for the area. These three conditions are typical of El Salvador and have been taken as the basis for this present discussion.

## Pre-drying

The coffee, already washed, is driven out onto the terraces by a stream of clean water. On arrival, the coffee is separated from the water that has been propelling it along, spread out in layers 15-20 cm high, arranged in long lines and turned periodically during the day in order to eliminate the moisture gently (Choussy, 1940).
    The coffee that is spread out on the terraces contains about 60 per cent water (kg water/kg dried coffee). After some three days in the sun, the water content has

dropped to about 45 per cent and the pre-drying process is considered complete. The coffee can remain undisturbed on the terraces until the water content has been reduced to 12 per cent or it can be dried in special machinery (Núñez, 1971) more rapidly. (There are machines for pre-drying coffee in which the interstitial water of the bean is removed first and then the superficial water. But these machines are relatively expensive and are not yet in general use.)

## Drying

Forty years have passed since the initial introduction of machinery specially designed to dry coffee (Choussy, 1940; Rehm, 1937). Only in the last twenty years has their use become common, having successfully overcome the obstacles that existed and allayed the fears that these dryers could be harmful.

The machinery used in El Salvador to dry coffee functions by weight and generally consists of one of two models: the horizontal and the vertical (Figure 7.3). The principle behind the operation is essentially the same in both models: a layer of air is crossed continually by a current of air heated to 60°C (only at the end of the process is cold air circulated to cool the coffee bean). In one dryer, the coffee forms a thick layer curving horizontally; in the other, the layer of coffee is thin, vertical and can be curved or flat. In order to ensure that the coffee is uniformly dried, it is mixed during the drying; in the case of horizontal dryers, the coffee is carried up to the top and deposited there in a continual process.

## SOURCES OF ENERGY IN USE

THE VAPOUR-BASED HEATER is one of the machines that has been in use for the longest time. Its advantage is that it can be adjusted easily to suit the varying conditions of the incoming air. Besides, the air that goes into the dryer is not mixed with any substance that could contaminate it while being heated. There are also machines that use a heat exchanger run on *Calor* gas/gas. On one side flow the gases resulting from combustion, generally from the skin of the coffee beans; on the other side flows the air that is going into the dryer.

However, the most favoured machine, particularly in the last few years, is the burner of combustible liquid (diesel). This is a small, compact machine, with a relatively low initial cost, in which the gases from combustion are mixed with the rest of the air that is going to the dryer. A thermostat regulates the fuel injection and a supervisor corrects whatever imbalance in combustion might occur.

Figure 7.4 shows a comparison of these three methods. The burner of combustible diesel transmits energy in the most direct form into the air it is drying.

**Figure 7.3  Diagram of Coffee Dryers most frequently used in El Salvador**

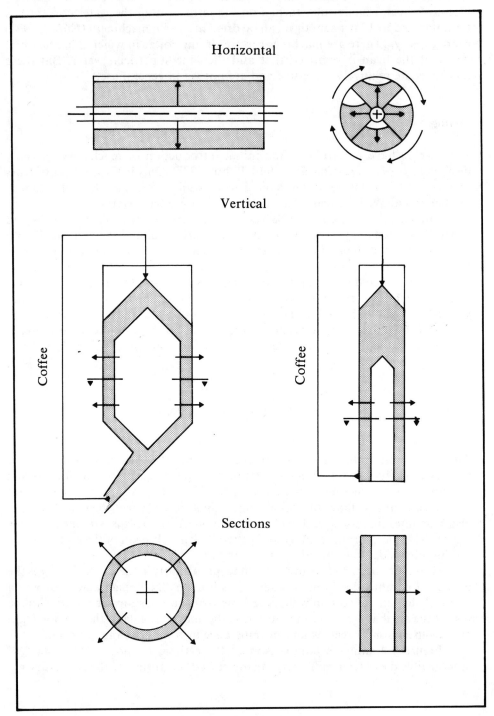

CASE STUDIES: DRYING COFFEE WITH RENEWABLE ENERGY 147

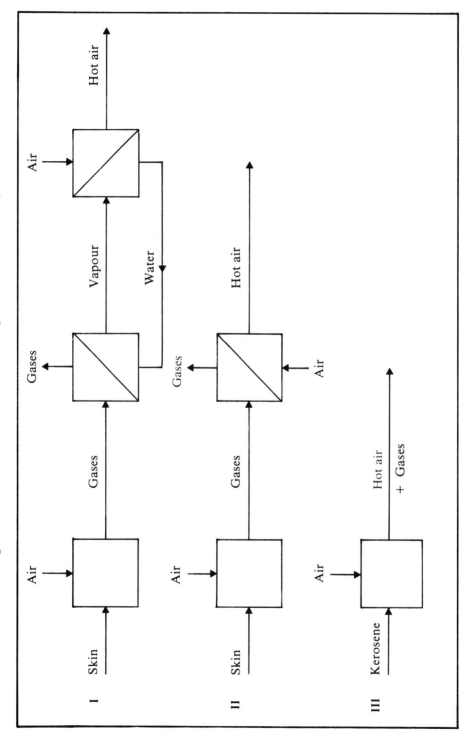

Figure 7.4  Methods in use for Heating the Air in Coffee Dryers

## Non-conventional Sources of Energy

The methods of heating air that use non-conventional sources of energy can be classified according to the number of heat exchangers used. This classification is based on the fact that the efficiency of the heat exchangers should be doubled when these apparati are connected 'one on top of the other'. Among the methods that do not use heat exchangers is the direct heating of air by solar rays. But this is limited to areas with little cloud formation and is of little interest in El Salvador where coffee plantations are situated in low to high cloud areas. The use of wind power could give rise to a system that does not require heat exchangers. It would entail generating a perpetual current that would be used to obtain oxygen and hydrogen from the water by way of electrolysis. These two gases would combine again in the burner that heats the air before it goes to the dryer. It might also give rise to another form of heat exchanger since the current generated by the wind could heat the air by means of electrical resistances. In order to use this wind power, however, plants that are fairly expensive to install are required (Portola Institute, 1974). These cannot compete in price with combustible materials and with other methods for drying coffee. It is possible, however, that their cost will drop in the future and that they will once more be of interest to coffee growers in areas subject to strong winds.

Pyrolysis is a process that permits the conversion of the skins of the coffee into a low-grade gas (Eggen and Kraatz, 1976). The anaerobic digestion of the coffee pulp permits the generation of methane (Black, 1960). Both combustible gases can be used to heat the air that goes to the dryers without the use of heat exchangers.

Among the methods using heat exchangers is the method of heating air by means of a solar collector. This absorbs the solar radiation and transfers it, in the form of heat, to the current of air going to the dryer. There are two heat exchangers in action in the case of the solar collector that uses an intermediary fluid to transmit the heat from the collector to the air heater. But in those cases where the solar collector uses an accumulation of hot fluid, there are three heat exchangers: the transfer of heat taking place in the collector, the storage tank and the air heater.

## USE OF WASTE PRODUCTS

THE WASTE PRODUCTS from coffee-processing have always been of considerable interest (Aguirre, 1966). This is due to the fact that they have no commercial value, are produced in considerable quantities and are made-up of a type of raw material that has never been successfully used. In order to analyze its utility, let us imagine a coffee plantation that produces 908 metric tonnes of peeled coffee (golden coffee) in a harvest lasting 100 days. A dryer that can produce 9,080 kg per day is sufficient for their needs, that is 13,620 kg in 36 hours. From this total of hours, which seems large in comparison with some experiments (Ives, 1955), thirty hours are necessary for heating the air, the remainder for loading, unloading and cooling the coffee.

A dryer with a capacity of 13,620 kg uses, according to measurements taken, 10.22 cubic metres per second of air and requires something equivalent to 411 kW to heat the air from 20°C to 60°C. The heat necessary for thirty hours of heating would use 1,154 cubic metres of diesel, at the value of 365.83 Colones. Thus, in order to dry an entire harvest of this size with diesel, one would have to spend 24,389 Colones on fuel alone.

## Skin

Every 120 kg of unpeeled coffee produces 20 kg of skin and 100 kg of golden coffee beans (Núñez, 1971). The imaginary coffee plantation, just taken as an example (producing 908 metric tonnes per annum of golden coffee), produces 227 metric tonnes per annum of skin. This quantity of skin (about 1,100 cubic metres in volume) can be stored in a warehouse 200 square metres in size, at a construction cost of some 35,000 Colones. The cost of the machinery to process this quantity of skin by pyrolysis in 100 days has been estimated at 20,000 Colones.

Fifty per cent of the skin is made up of cellulose. The combustion heat of this material is 18,600 kj/kg. If we assume that only two-thirds of the heat from its combustion is transferable to the air in the dryer, we discover that the volume of skin available (95 kg per hour) can produce a heat equivalent to 164 kW. Therefore, the skin could be substituted for the 48 per cent diesel required to heat the air in the dryer. These figures are based on the fact that pyrolysis produces 5,904 kWh in 36 hours and the dryer requires 12,330 kWh for 30 hours. The combustible diesel that it could replace is worth 11,707 Colones a year.

## Pulp

Every 100 kg of ripe coffee (grape) produces 40 kg of fresh pulp and 20 kg of golden coffee. A process using anaerobic digestion, with 12 days deduction process, produced 60 litres of methane, generated by 1 kg of fresh compressed pulp. For the digestion process, the pulp is mixed with an equal volume of water.

The coffee plantation imagined here processes 9.08 metric tonnes per day of golden coffee: 18.16 metric tonnes per day of fresh pulp (218 metric tonnes in 12 days) with a total volume of 1,090 cubic metres. Providing 20 per cent additional space, the pulp requires a tank 1,308 cubic metres which, if constructed in concrete, would cost 163,500 Colones. Lacking the precise figures, we have added an estimated 50 per cent of the above cost to cover the installation of extra fixtures in the plant (pumps, pipes, valves, controls, water washer, compressor). In this case, the initial cost of investment is 245,250 Colones. A volume of 18.16 metric tonnes per day of fresh pulp generates 1,089.6 cubic metres per day of methane (45.4 cubic metres per hour). This volume of methane with a combustion heat of 35,960 kj/$m^3$ can produce a heat equivalent to 453 kW (Hütte, 1955). If we take into account that the dryer uses 12,330 kWh in 30 hours and that the methane produces 16,308 kWh in 36 hours, we see that this process could provide 132 per cent of the total heat required to dry the coffee. The methane generated by the pulp can be substituted for the entire quantity of diesel and can produce an additional quantity worth 7,804 Colones.

**Use of Solar Energy**

In El Salvador, the overall solar radiation is equivalent to $0.654\,kW/m^2$. We shall imagine that one-third can be transferred to the air in the dryer (Stessels and Pridmann, 1972; Stessels, 1973). Under these conditions, 1,885 square metres is required to heat the air in a dryer with a capacity of 13,620 kg, from 20°C to 60°C. In order to dry this amount of coffee, the diesel burner must produce 12,330 kWh in 36 hours (30 net hours of heating). For every 24 hours, 8,220 kWh are required. Solar energy in this instance is capable of producing 3,288 kWh in 8 hours of sun. This means that the solar collector can replace the 40 per cent of diesel required to dry the coffee. This is equivalent to the value of 9,756 Colones per year.

The solar heater proposed for this work is made up of two sheets of corrugated, galvanized metal painted on the outside with black matte and placed one on top of the other. The two sheets are rested on the ground at a 25° angle and slope towards the south. The air going to the dryers is sucked in by ducts formed by the two sheets. The calculus that has to be adjusted to the acceptable fall in pressure also regulates the air speed in the ducts and the transfer of heat. We shall accept a low standard of efficiency intentionally, in order not to raise the cost or complicate the installation of the collector. An installation of this kind would mean an initial investment of 39,414 Colones inclusive of the collector sheets and 100 metres of piping from the collectors to the dryer.

## COMPARISON OF THE USES OF NON-CONVENTIONAL SOURCES OF ENERGY

FOR EACH ONE of the alternatives that have been discussed (pulp, skin and solar energy), we have calculated the annual cost, divided into fixed costs (depreciation and interest at 11 per cent per annum) and the annual cost of operation (manual labour, spare parts and the combustible diesel required). The working life of a concrete tank (pulp) has been suggested as 50 years, with a residual value of zero; that of a solar plant and the warehouse (skin) as 20 years, with a 10 per cent residual value, and a working life of 10 years with a 10 per cent residual value for other cases.

We emphasize the anaerobic digestion of pulp as an alternative. In this way, 76 per cent of the pulp produced is used to generate the energy necessary to dry the coffee. (As previously mentioned, the pulp from a coffee plantation is capable of producing 132 per cent of the heat required by the dryer.) All the calculations refer to actual prices in El Salvador. In other countries, prices — particularly those of

CASE STUDIES: DRYING COFFEE WITH RENEWABLE ENERGY 151

combustible diesel — may be noticeably different. We must consider, in addition, that it will be the unpredictable factors that will be of the greatest importance, emphasizing and relying on the ability of local engineers to design and construct the type of equipment required.

The costs for each alternative are presented independently in Tables 7.7 and 7.8. We have indicated the initial costs of investment, the value of energy generated and the percentage of diesel replaced (by these non-conventional sources of energy) in each alternative method. Table 7.9 compares the alternatives. We have taken into account that only pulp is capable of supplying the total heat requirement. (The other two alternatives require combustible diesel.) We have also included a column in which we have combined solar energy with pyrolysis and with combustible diesel.

Finally, two further aspects should be mentioned: the design of the dryers and the economizing of energy during the drying process. The amount of waste expelled by the air leaving the dryers is considerable (Tosello, 1946) and the time required to dry the coffee seems lengthy. There is a solution to both problems. The defects apparent in the machinery actually in use can be corrected by improving the transfer of hot air to the coffee bean (Chiquillo, 1977). And the consumption of fuel can be reduced by recirculating the air that leaves the dryer and by the use of heat recuperators. The air to be recirculated should be mixed with varying proportions of fresh air so that there would be a high percentage of recirculated air towards the end of the drying process and a very low percentage at the beginning. These methods of perfecting the equipment and the process are of fundamental importance, and will considerably increase the efficiency of the alternatives presented here.

Table 7.7  Annual Cost of certain Non-conventional Sources of Energy used in the Drying of Coffee (in colones)

|  | Diesel | 0.76 Pulp | Skin | Solar |
|---|---|---|---|---|
| Depreciation | 650 | 8,100 | 3,400 | 1,800 |
| Interest | 450 | 10,600 | 3,300 | 2,400 |
| Fixed costs | 1,100 | 18,700 | 6,700 | 4,200 |
| Cost of labour | 1,700 | 5,700 | 2,500 | 2,500 |
| Spare parts | 300 | 5,400 | 3,000 | 2,400 |
| Fuel | 24,400 | — | 12,700 | 14,600 |
| Operation | 26,400 | 11,100 | 18,200 | 19,500 |
| Annual cost | 27,500 | 29,800 | 24,900 | 23,700 |

**Table 7.8** Initial Investment, Value of Energy generated and Percentage of Diesel substituted in the use of Non-conventional Sources of Energy for the Drying of Coffee

|  | Diesel | 0.76 Pulp | Skin | Solar |
|---|---|---|---|---|
| Initial investment (colones) | 7,000 | 186,000 | 55,000 | 39,000 |
| Value of the energy generated (colones) | — | 24,400 | 11,700 | 9,800 |
| Percentage of diesel substituted | — | 100 | 48 | 40 |

**Table 7.9** Annual Cost of Using certain Non-conventional Sources of Energy to Dry Coffee (including combined sources of energy)

|  | Diesel | 0.76 pulp | Skin and Diesel | Solar and Diesel | Skin, Solar and Diesel |
|---|---|---|---|---|---|
| Depreciation | 650 | 8,100 | 4,050 | 2,450 | 5,350 |
| Interest | 450 | 10,600 | 3,750 | 2,850 | 6,150 |
| Fixed costs | 1,100 | 18,700 | 7,800 | 5,300 | 12,000 |
| Cost of labour | 1,700 | 5,700 | 4,200 | 4,200 | 6,700 |
| Spare parts | 300 | 5,400 | 3,300 | 2,700 | 5,700 |
| Fuel | 24,400 | — | 12,700 | 14,600 | 2,900 |
| Operation | 26,400 | 11,100 | 20,200 | 21,500 | 15,300 |
| Annual cost (colones) | 27,500 | 29,800 | 28,000 | 26,800 | 27,300 |

### References

Aguirre, B.F. (1966) *La Utilizaçion Industrial del Grano de Café y sus sub-productos.* Investigaçiónes Tecnológicas del ICAITI, No.1, Instituto Centroamericano de Investigaçíon y Tecnologia Industrial, Guatemala.

Black, H.H. (1960) *Engineering Studies of Coffee Mill Wastes in El Salvador.* Report to International Co-operation Administration, USA Operations Mission to El Salvador, US Dept. of Health, Education and Welfare, Public Health Service, Robert A. Taft Sanitary Engineering Centre, Cincinnati, Ohio.

Bosnjakovic, F. (1960) *Technische Thermodynamik,* Vol. 2, Th. Steinkopff Verlag, Dresden/Leipzig.

Coste, R. (1960) *Cafétos y Cafés en el mundo,* Vol. 2, Edif. Hachette, Paris.

Chiquillo, A. A. (1977) *Mejoras a las secadoras verticales de Café*. Informe interno, inedito. Compañia Salvadoreña de Café, San Salvador.

Choussy, F. (1940) *Estudios Tecnicos de la Secada del Café: El Café de El Salvador*. Revista de la Asociacion. Cafetalera de El Salvador, Vol. 8, pp. 585-620, 641-670, 724-740, 790-815.

Eggen, A.C.W. and Kraatz, R. (1976) *Gasification of Solid Wastes in Fixed Beds*. Mechanical Engineering, Vol. 98, No. 7, pp. 24-29.

Hütte, J. (1955) Vol. l. W. Ernst und Sohn Verlag, Berlin.

Ives, N.C. (1955) *Estudios sobre Secamiento de Cafe*. Turrialba, Vol. 5, Nos. 1-2, pp. 17-25.

Núñez, L.R.A. (1971) *Analisis de los Procesos del Beneficiado del Cafe*. Tesis, Facultad de Ingenieria y Arquitectura, Universidad de El Salvador, San Salvador.

Portola Institute (1974) *Energy Primer: Solar, Water, Wind and Biofuels*. Menlo Park, California.

Rehm, W. (1937) *El Secamiento del Café con los Secadores MIAG: El Café de El Salvador*. Revista de la Asociacion Cafetalera de El Salvador, Vol. 5, pp. 724-729.

Stessels, L. (1973) *Utilisation de l'Energie Solaire pour la conservation du Café en Region Tropical Humide:* Project de magasin industriel de stockage. Café, Cacao, Thé, Vol. 17, No. 2, pp. 142-158.

Stessels, L. and Pridmann M. (1972) *Utilisation de l'Energie Solaire pour la conservation du Café en Region Tropical Humide:* Café, Cacao, Thé, Vol. 16, No. 2, pp. 135-148.

Tosello, A. (1946) *Ensaios sobre a Secagem dos Produtos Agricolas Pelo arguente*. Bragantia, Boletim Técnico da Divisao de Experimentaçao e Pesquisas, Instituto Agronómico, Campinas, Vol. 6, No. 2, pp. 39-108.

# IX Development and Diffusion of Clay Stoves for Firewood Economy in Rural Areas of Guatemala

*Roberto Cáceres*
CEMAT for OLADE

WOOD AND FORESTRY resources are amongst the most important renewable sources of energy for the Third World. Firewood is the primary source of energy for more than one-third of the world's population. Rapid extinction of forestry resources, detected in almost every developing country, opens a new energy crisis, meaning extinction after 20-30 years of basic energy for cooking in the popular groups. Apart from a negative influence on the satisfaction of the population's basic needs, increasing deforestation is affecting the basis of food production by soil exhaustion, desertification, floods and climate variations (Eckholm, 1979).

Centro Mesoamericano de Tecnologia Apropriada (CEMAT) is a non-governmental organization, founded after the 1976 earthquake in Guatemala for national reconstruction duties. It intends to contribute towards this huge challenge of developing and diffusing appropriate technologies for firewood economy and bring out alternative sources of energy and fertilization. The following technologies have been selected for permanent evaluation, improvement and adaptation: (1) stoves constructed with earthen materials that economize more than 30 per cent of firewood, compared to the archaic way of cooking; (2) use of biogas, produced in the rural areas through biomass digestors, within the reach of rural communities.

In this discussion, attention will be focussed on the experience of Mesoamerica, specifically in Guatemala's Occidental High Plateau, where the development and diffusion of earthen stoves to economize on firewood was carried out.

## DEFORESTATION AND WASTE OF RENEWABLE PRIMARY ENERGY RESOURCES

DURING THE LAST two decades, a process of increasing deforestation began due to the increasing demand for wood which developed faster than new trees could grow (Makhijani and Poole, 1975). This extinction of forestry resources is part of a general process involving the degradation of natural resources, which is closely related to accelerated consumption, due to greedy and uncontrolled models of development, in addition to population growth.

In Guatemala, 60 per cent of the land was forested in 1950. In 1975, forest land had been reduced to 36 per cent. Thus, in 25 years, 44.6 per cent of forests were lost at an annual rate of 2 per cent. Estimated losses over this period amount to 2,400 million quetzales (1 quetzale =US$ 1). If to this we add indirect effects of accelerated deforestation, we have a loss of hydric resources at a reposition cost of 200 million quetzales and a loss of 32 per cent of the productive capacity of agricultural land, which gives an annual loss of 420 million quetzales. Farmlands not being retained are lost to the sea.

Apart from these elements, there are still other causes in Guatemala that have assisted in deforestation: the need for wood for house reconstruction after the earthquake, the presence of the pine-mite plague and expansion of cattle-raising and cotton. Wood consumption in rural areas is greater for cooking purposes. In Guatemala, a peasant family uses approximately one cubic metre of firewood each month (Evans, 1979). For the existing 600,000 peasant families in the country, this represents an annual consumption of 7.2 million cubic metres of wood. Guatemala, which etymologically means 'country of forests', is getting close to a situation where wood is becoming very scarce and, in the next decade, a similar condition may exist as in some Caribbean countries, the African Sahel and the dense Asian regions, where firewood is a rare commodity. If many Latin American countries with an abundance of forestry resources maintain present consumption rates, they too will have complete deforestation in a 20-30 year period.

| Country | Per cent Forest Coverage | Estimated Period for Complete Deforestation (Years) |
|---|---|---|
| Brazil | 67 | 30 |
| Colombia | 74 | 30 |
| Costa Rica | 38 | 22 |
| Guatemala | 36 | 25 |

Source: Ferraté and Klussmann, 1978; Parham, 1978.

Facing this fact, Guatemala's forestry situation has been declared an emergency. A law for support of reforestation went into effect and duties at ministerial level have been assigned. An evaluation of this campaign would be premature but it is a fact that trees planted this year will fructify, if looked after carefully, after only five to eight years. Because of this, all plans that tend to decrease rates of consumption of firewood in rural areas are considered important.

Firewood consumption is only one problem; other important issues to be tackled are the extension of the agricultural frontiers and the cutting of wood to furnish agricultural and cattle-raising lands.

**Archaic Cooking Methods and Basic Needs for Firewood**

The majority of families in rural areas of Guatemala, who are the largest group in the country, basically use firewood for cooking their food. The most common technique is the archaic method of open fire cooking, which was adequate when there was a relative abundance of firewood. This archaic way of cooking consists of using three stones over which a pot or *comal* (a large clay dish for cooking corn *tortillas,* the population's basic nourishment) is set; the firewood is laid below to produce open-air fire. This method of cooking has existed for centuries in the rural Mesoamerican area. The open fire allows atmospheric warming, the smoke protects straw roofs from animals that tend to shorten its serviceable life and, in certain zones, the smoke drives away mosquitoes. To all these facts we have to add the element of tradition, whereby women sit on the ground to prepare and cook food.

In this sense then, access to firewood is a basic need for peasant families. Normally, an average peasant family needs one *tarea* (approximately one cubic metre of firewood, cut and heaped) per month. Where there still exist accessible neighbouring woods, families send some of their members for daily provisions, which they cut after their working day. Where wood is scarce, or if working sites are far away, it is necessary to buy firewood at a price of between 8-20 quetzales which represents one-fourth of the family's monthly income.

## DEVELOPMENT AND IMPROVEMENT OF EFFICIENT STOVES MADE WITH LOCAL MATERIALS

IN ORDER to offer an alternative to open-fire cooking, technologies which would be appropriate to the economic, social and ecological conditions of the communities were examined, techniques that could be incorporated into the culture of the groups. It was necessary to use local workmanship and simple techniques to be carried out at a family or small group level at a low cost. It was also important to allow generation of self-determination in the diffusion process (Salinas, 1978; Reddy, 1978).

The design was based on the experiences of many developing countries— especially India, Egypt, Ghana and Indonesia—where the basic principles of the *Choola* of Hyderabad model were improved (Raju, 1953). The process began in 1953 and probably its most promising results were gained in the Lorena Stove model, developed in Guatemala in 1976 due to needs generated by the reconstruction programme after the earthquake (Evans, 1979).

The essential features of this group of stoves made with earthen materials are the following:
(1) They are constructed from a base of clayish soil that preserves heat, blended with sand, excrement, straw and other materials that do not allow the splitting of the mixture.
(2) Basic knowledge of work with soil is required, which any peasant using simple tools can accomplish.
(3) Firewood is used, together with other elements such as coal, cane and *olotes.*

At the end of the 1950s, a campaign was devised in Guatemala to introduce stoves built with cheap, prefabricated materials. These materials would be freely distributed; the user would furnish the wooden base. Because of a lack of follow-up and motivation of the rural communities, this programme failed. Its massive introduction at that time would possibly have prepared the way for present developments.

Open-air cooking is inefficient from the energy point of view. Only 5-10 per cent of the potential fuel energy of firewood is used in cooking. Moreover, open-air fire practice has other considerable disadvantages:
(1) smoke, which spreads through the house atmosphere, is a constant discomfort and cause of many ocular and broncho-pulmonary disorders;
(2) possible burns for children, due to proximity and poor protection from fire;
(3) contact with the floor and consequently with animals and dust make kitchen surroundings unhygienic;
(4) due to limited control over fire gradation, there is more work.

For peasant families with more resources than the average, the *poyo* (a brick or stone stove, or stones with an iron cast plate to settle pots) is traditional. The problem is its price: 60-100 quetzales. It overcomes the presence of smoke but it uses up more wood than open fires. Due to its efficiency, produced by the interaction of its firebox and air draught, the *poyo* gives a good performance that makes possible an approximate 20-50 per cent firewood economy, compared to open fire. It also uses a chimney that permits smoke evacuation and generation of air draught.

**Lorena Stove**

The present design of this stove was developed in Choqui Experimental Station in Quetzaltenango, Guatemala, towards the end of 1976. The first promoters who began to diffuse the technique in the Guatemala Occidental High Plateau were trained there. To build the Lorena stove (Lorena is a word formed by the compound *lodo* and *arena,* which means mud and sand), an adobe or block base must be built. It is raised 40 centimetres on a solid block of mud and sand. After this block has dried, holes for the pots must be made. Relatively long tunnels are pierced, with different twirls to secure heat absorption, both from the mud and sand, as from the pots. As smoke is expelled, it leaves heat in this system, allowing up to four openings for placing cooking vessels. Closer to the chimney, temperature decreases; finally, there is permanent hot-water storage (Figure 7.5). The heat of the firebox cooks the first pot. Hot air circulating towards the chimney crosses over the tunnel system until it flows into the chimney through the air draught and, in its course, heats the other pots adjusted in the openings.

A well-constructed model can retain up to 90 per cent of fuel heat. Besides, this mud and sand stove possesses a series of air flow regulators that provide greater efficiency, smoke control and enormous independence to the cook in relation to the fire. The fact that they are constructed with mud gives them a domestic form that unites harmoniously with the peasants' native culture. When coated in cement, these stoves can even be displayed in an urban kitchen.

**Diffusion Programme for the High Plateau**

The goals of the diffusion programme for the Lorena stove over the years 1979-1980 (first stage of massive diffusion) are the following (Cáceres, 1978):

(1) **Educational Goals**
   (a) Qualification of stove builders: the purpose is to train stove technicians

Figure 7.5    The Lorena Stove Model

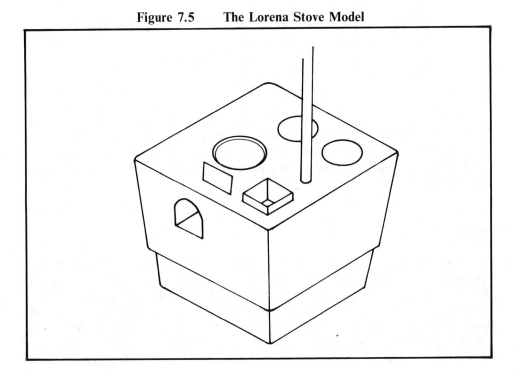

from poor peasant people in the areas of Lake Atitlan, Chimaltenango, Quetzeltenango and Totonicapan;
(b) Short qualification courses: specialized short courses are granted to teach stove builders, charging a small fee for expenses, according to their financial capabilities;
(c) Short course follow-ups: follow-up training for participants is devised through trips to their towns to perceive their experience in stove construction or counselling for improvements to stoves with defects. Each trimester, builders meet to share their experiences and pursue specialized training;
(d) Professorship: expert stove constructors are given a diploma; their associative organization is thus stimulated.

(2) **Economic Goals**
(a) Firewood economy: to decrease by 50 per cent the use of firewood and improve cooking of food;
(b) Employment generation: stove construction is a remunerative activity that generates employment. Increasing stove construction increases employment rates;
(c) Generation of induced local wages: construction of stoves with local materials and local workmanship originates small indirect revenues for chimney manufacturers, sand or sometimes mud carriers and merchants of cheap kitchen utensils;
(d) Support to domestic work: women's work is favoured because stoves permit finer cookery, more food and less water in less time, increase domestic work productivity and decrease the amount of working time, allowing women a greater amount of free time.

(3) **Environmental Improvement**
(a) Improvement of the kitchen atmosphere: the purpose is to reduce smoke from peasant kitchens, which creates various broncho-pulmonary illnesses;
(b) Increase of food hygiene: using the stove, food is generally kept off the ground, reducing contamination. Boiling water produced by the stove diminishes to a reasonable extent water contamination dangers;
(c) Forestry erosion decrease: by means of the Lorena stove's massive use, collective use of firewood is reduced, decreasing forestry wearing, which is one of the Occidental High Plateau's most serious problems.

## Diffusion in Guatemala's High Plateau, a Region of Minifundium Peasants

The Occidental High Plateau of Guatemala is inhabited mostly by peasants of Indian origin who possess small parcels of land. Their annual income is scarce, not more than 200 quetzales per year. The energy crisis, earthquake effects, lack of farming lands and soil weariness have impoverished these peasant groups of the

High Plateau. Relative firewood scarcity, which has begun to be felt directly, affects one-fourth of their income and their quality of life. The experience to be analyzed takes place in five communities of the High Plateau, with emphasis on one located around Lake Atitlan.

This community is relatively dense, having more than 700 inhabitants per square kilometre of farmland; population growth rate is 2.4 per cent (Salinas and Cáceres, 1978). Firewood consumption in this community is approximately 12,000 cubic metres per year. Consequently, deforestation is accelerated.

Climate variations that affect agriculture are observed. Commercial cultures, such as coffee and onions, have been developed in the last years, harming subsistence crops. Inhabitants of this community are famous for having been the first to introduce coffee and they have made great efforts to educate young generations. This community is where the Lorena stove has been most successfully introduced. To an extent, it has become part of the local peasant culture.

Actually, some 200 families possess a Lorena stove, approximately 5 per cent of the total, which we consider the minimum to be able to assume acceptance within the local culture. Development and diffusion of Lorena stoves in this community has opened new possibilities of work and organization. Presently, ten expert stove builders are dedicated to stove construction in this and adjacent communities. What are the factors that allowed this?

> (1) the existence of enthusiastic promoters, who have constructed stoves, first in their own homes and then have taught the rest;
>
> (2) follow-up and technical assistance given to the promoters who have been initiators;
>
> (3) the efforts of the group formed who have organized themselves, on their own initiative, into small technical groups.

In the High Plateau, three other communities have 1-2 per cent stove builders in their families, and four communities between 0.5-1 per cent. Other stove types have no significance from a massive diffusion point of view.

## PRELIMINARY EVALUATION OF THE LORENA STOVE

CHOQUI AND CEMAT Experimental Stations, who work jointly in the diffusion of Lorena stoves in the High Plateau, have done preliminary evaluations. Studying 64 families selected at random, the following observations were made (Estación Experimental Choqui, 1979):

|  | Percentage |
|---|---|
| Number of stoves that perform daily | 91 |
| Stoves badly constructed or placed | 9 |
| Stoves built by promoters | 26 |
| Stoves built by participants | 74 |
| Fissured stoves* | 73 |
| Stoves that do not produce smoke | 38 |
| Stoves that produce little smoke | 58 |
| Stoves that heat quickly | 93 |
| Satisfied owners | 95 |
| Families who use less firewood | 90 |

The principal conveniences expressed by the people interested in learning this technique are the saving of firewood and the low cost. After the qualification courses (which last two to three days), the trend is that at least one fifth of the participants are interested in teaching others.

There is also a tendency to change the stove shape. This was viewed initially by the course engineers as a negative trend to be corrected. Later, we acknowledged this as positive progress, meaning the adaptation of stove construction to each peasant family's specific needs and environment (see list of design modifications). At first, there was a tendency to build stoves in characteristic sites, but generally they are artificial; there is no person who cooks permanently and maintains the stoves adequately. In this sense, we saw the important role of women in stove diffusion (Cáceres, 1979). In colder places, the women complained that the stoves did not produce enough heat to warm rooms. This could be a long-term obstacle in the regions we are studying.

Assessing this first stage, we consider that Lorena stoves have endured the trial; the diffusion methods are producing results. The next stage of massive diffusion can only be accomplished with the assistance of development institutions and public participation in the diffusion of this technology. At an international level, there actually exists great concern over world diffusion of the Lorena stove and we are especially interested in forming series of groups to work in applied technology to diffuse, adapt and evaluate it as an instrument to save firewood for peasant families of developing countries (CEMAT, 1979).

**Design Modifications of the Lorena Stove**

The following features can be modified:
  (1) size and shape;
  (2) quantity and size of the holes for pots;
  (3) size of the firebox;

---

*High fissure rates appeared in the most humid zones, where drying takes more time. Stove builders perforated tunnels before mixtures had dried. Blends were inadequate and, more important, technical follow-up was practically nonexistent.

(4) diameter and length of the chimney;
(5) base materials;
(6) number of tunnels;
(7) angles of tunnels;
(8) placing of the water heater;
(9) its function as stove or heater.

The following specifications of features cannot be easily modified:
(1) stoves always need a chimney;
(2) smoke escapes must not exist;
(3) air entry points to the firebox should not be allowed;
(4) hatches must be hermetically locked;
(5) there should be at least two hatches, one in front of the firebox, the other facing the chimney;
(6) there must be a hot swirl against each pot base;
(7) the firebox should not be located too near to either side of the stove;
(8) the opening of the firebox hatch should be at a proper distance from its entrance;
(9) tunnels should be as large as possible and originate behind the firebox.

## References

Appropriate Technology International (1978) *Sand and Clay Stove Dissemination as Catalyst in Community Development.* Washington, DC.

Arnold, J.E.M. and Jongma, Jules (1978) *Fuelwood and Charcoal in Developing Countries.* FAO Rome, pp. 2,3,5,6,7.

Cáceres, Roberto (1978) *Proyecto de Difusión de Estufas de Lorena en el Altiplano de Guatemala.* CEMAT, Guatemala. (Unpublished).

Cáceres, Roberto (1979) *Importance de l'Education dans l'Appropiation Villageoise des Techniques de Base.* CEMAT/NGO Forum. Congreso Mundial sobre Ciencia y Tecnología para el Desarrollo, Vienna.

CEMAT (1979) Periódico RED.

Eckholm, Erik (1975) *The Other Energy Crisis: Firewood.* Worldwatch Paper 1, Worldwatch Institute, Washington DC, pp. 6, 7, 8, 9, 10, 15, 16.

Eckholm, Erik (1979) *Planting for the Future: Forestry for Human Needs.* Worldwatch Paper 26, Worldwatch Institute, Washington, DC, pp. 1, 19, 21, 29, 40, 44, 47, 56.

Estación Experimental Choqui (1979) *Evaluación Preliminar del Proyecto Estufas de Lorena.* Quetzaltenango.

Evans, Ianto (1979) *Lorena Owner-Built Stoves.* The Appropriate Technology Project of Volunteers in Asia, Stanford, California.

Ferraté, Luis A. and Klussmann, Evelyn (1978) *Terramoto y Ecocidio.* Memorias del Simposio Internacional sobre el Terremoto de Guatemala del 4 de Febrero de 1976 y el Proceso de Reconstrucción, Tomo II, Guatemala.

Goldemberg, J. and Brown, R.I. (1979) *Cooking Stoves: The State of the Art.* Institute of Physics, Universtiy of Sao Paolo, Brazil, pp.2, 8, 11, 13.

Makhijani, Arjun and Poole, Alan (1975) *Energy Agriculture in the Third World.* Ballinger Publishing Co., Cambridge, Mass., pp. 68,69,109.

Morgan, Robert and Icerman, Larry (1979) *Appropriate Technology for Renewable Resource Utilization.* Washington University, Washington, DC.

Parham, Walter (1978) *LDC Deforestation Problem - A Preliminary Data Base (Draft).* Dept. of State, Office of Science and Technology, Washington, DC.

Raju, S.P. (1953) *Smokeless Kitchens for the Millions.* The Christian Literature Society, Madras, India, pp. 1, 37.

Rao, E.G.K. (1962) *Improving the Domestic Choola.* Reprint from Indian Farming.

Reddy, Amulya Kumar N. (1978) *Energy Options for the Third World.* Bulletin of Atomic Scientists, Vol.34, No.5, pp. 28-33.

Revelle, Roger (1978) Requirements for Energy in the Rural Areas of Developing Countries. In *Renewable Energy Resources and Rural Applications in the Developing World.* Norman L. Brown (Ed.), Westview Press, Boulder, Co., pp. 13,24.

Salinas, Bertha (1978) *Tecnología Apropriada: Concepto, Aplicación y Estrategias.* CEMAT/UNESCO, Bogota, Colombia.

Salinas, Bertha and Cáceres, Roberto (1978) *Tecnología Apropriada para Suministro de Agua y Eliminación de Desechos: Un Estudio de Caso de una Comunidad del Lago.* Banco Mundial/CEMAT, Guatemala.

# CHAPTER 8

# Conclusions and Recommendations

## CONCLUSIONS

(1) Latin America possesses sufficient human resources to tackle more technologically the problems with regard to the exploitation of less sophisticated renewable energy sources.

(2) All countries of the region possess installed industrial capacity to manufacture equipment of intermediate-level technological sophistication; however, none of them (except for Brazil in the case of alcohol) is capable of manufacturing automatically equipment of greater technological complexity.

(3) The conditions of the region are such that the equipment required for the exploitation of less technologically sophisticated and renewable energy sources may become commercially viable.

(4) No significant efforts at the institutional and planning levels have been detected in the countries of the region (except Brazil with the alcohol programme) of assigning priority to the study, development and use of renewable sources of energy.

(5) Solar and wind energy technologies share more possibilities of penetration in the future.

(6) The problems that interfere with the investigations in the field of renewable energy sources are: lack of economic resources, instruments, equipment, laboratories, specific bibliography and exchange at a regional level of experience, technologies and personnel.

(7) Some countries even lack information and do not possess an institutional apparatus enabling them to face the task of performing an evaluation of resources concerning renewable energy sources.

(8) The relative degree of development of the countries in the region and their average consumption of energy show considerable diversity, as does the availability of energy resources. Consequently, the problems at hand are also of a diverse nature and no solutions may be applied uniformly to all.

(9) The predominant style of development in Latin America is founded on abundant and low-price oil; that is an imitation of the prevailing style in the industrialized countries of the West and Japan. This situation has made it difficult for the countries of the region to finance their foreign debts and to solve their balance of payments problems. Moreover, this style of development has not enabled the majority sectors of the population to participate significantly in the fruits of the economic growth achieved (since these sectors have barely surpassed the subsistence level of consumption) and, on the contrary, has further accentuated the divergences among social sectors.

(10) The present world and regional energy situations preclude considering development for Latin America based on the 'oil-intensive' model. Other approaches to development must be adopted and attention must be given to the use of existing technology that will lead to greater use of energy resources that have not yet been exploited because of abundance of oil.

(11) Intensive industrial exploitation of forests and the use of wood as the sole energy resource by vast numbers of the rural inhabitants in Latin America have brought about serious environmental problems, such as increased soil erosion and changes in microclimates.

(12) Hydroelectric power constitutes a great energy potential in the region, both with regard to large power stations and to the exploitation of small waterfalls and water flows. So far, sufficient study of this resource has not been carried out.

(13) An increase in the availability and consumption of energy as may be brought about in many instances by the use of renewable sources of energy would result in significant improvement of the quality of life of the inhabitants of the Latin American rural sector.

(14) It is important to bear in mind that the possible adoption of technologies for using renewable sources of energy does not imply replacement of technologies employed with regard to conventional energy sources. Rather, technological pluralism will make it possible to use both conventional and renewable sources by means of either simple or sophisticated technologies compatible with the environment and development of the areas in which they are used.

## RECOMMENDATIONS

(1) The encouragement of regional and subregional organizations dealing with energy problems to attach greater importance to, and include in their programmes, the use of renewable sources of energy. In addition, the promotion of special credit lines in international financing organizations for the use of such resources.

(2) Amendment of present legislation on the use of renewable sources of energy and the establishment of institutional machinery with effective power to formulate and implement national and regional energy policies that explicitly envisage the role such sources should play in the harmonious development of the countries of the region. Such legislation should promote technological pluralism in the use of the sum total of the region's energy resources.

(3) The establishment of the institutional (or legal) machinery to review transfer of technology contracts with regard to energy in order to avoid the inclusion of clauses that may prove detrimental to autonomous national development in this field.

(4) The establishment of national or regional scientific and technological research centres on the use of renewable energy sources that will make it possible to train people and avoid the creation of new and total technological dependence of the countries of the region on the most developed countries. Such centres should, *inter alia*:
- (a) Set up research programmes in accordance with the needs and plans of the countries in the region and organize specialized courses, seminars, study trips abroad and so forth to provide sound training for native researchers or others who work in institutions in Latin American countries.
- (b) Set up information systems that will inform researchers and producers on progress achieved with regard to the use of renewable energy resources in other countries of the region and outside the region itself.
- (c) Promote the constitution of interdisciplinary research groups in the use and impact of equipment for using renewable sources of energy. Such groups should study the principal cultural patterns of potential users in order to avoid problems of rejection or acculturation.
- (d) Carry out feasibility studies to select the technologies most appropriate to social and economic realities and most likely to have a positive impact on the economy of the countries or regions in question, as well as to improve the quality of life of the inhabitants.
- (e) Establish common standards for measuring the quantity and quality of renewable energy resources with a view to achieving homogeneity and making comparisons among countries to facilitate the exchange of experiences and joint research efforts.

(5) Formulation on the national level of various specific financing mechanisms for producers, distributors and users of equipment using renewable energy resources.

(6) Among the technologies of general interest, some are already sufficiently developed and have entered their commercial phase. Such is the case with respect to solar water heaters and small wind generators of less than one kW power. The

establishment of appropriate machinery is recommended to initiate their manufacture on an industrial scale and their dissemination in regions where they may be employed.

(7) Among other technologies requiring additional research and developments are:
- (a) *Hydraulic microturbine systems with electric generators:* the great hydraulic potential of most of the countries of Latin America with regard to uses inferior to 100 kW justifies the development of systems of this type. At present, additional technological effort could lead to more efficient systems which, with appropriate materials, could be designed for mass production.
- (b) *Low-power photovoltaic generators (less than one kW):* the reductions foreseen in the production costs of solar cells give rise to the belief that their use will increase considerably in the near future. Solar electric power generators have significant applications in rural areas as, for example, in the pumping of water for domestic or irrigation purposes and in communications.
- (c) *Medium-power wind generators:* in Latin American rural areas—particularly in regions favoured by wind resources—there is a need for equipment with power of from one to ten kW to provide electric power for a large number of agricultural tasks, especially with regard to the processing of agricultural products and food preservation. These needs can be covered by windmills designed for this purpose.
- (d) *Power units of less than 200 kW derived from non-fermented plant material:* these units would be used as stationary energy sources in agricultural operations. The development of alternative systems is also necessary; for example, steam engines driven by direct combustion of wood and agricultural wastes; internal combustion motors driven by the gas generated in gasogenes by the burning of charcoal; and the use of pyrolytic fuels. The development of such sound systems, easy to operate and maintain, is of great interest to many regions of Latin America.
- (e) *Biogas plants:* the development of small-scale systems and of methods for operating them with common raw materials (grain chaff, animal excrement and the like). Experience is varied in this subject area and it would be appropriate to systematize its development as a means of obtaining inexpensive and easy-to-operate equipment.